Fourier Transform Rheology on Dispersions based on Newtonian Fluids

Zur Erlangung des akademischen Grades eines
DOKTORS DER NATURWISSENSCHAFTEN
(Dr. rer. nat.)
Fakultät für Chemie und Biowissenschaften

Karlsruher Institut für Technologie (KIT) - Universitätsbereich
genehmigte
DISSERTATION
von
Kathrin Reinheimer
aus
Wittlich

Dekan:	Professor Dr. M. Bastmeyer
Referent:	Professor Dr. M. Wilhelm
Korreferent:	Professor Dr. C. Barner-Kowollik
Tag der mündlichen Prüfung:	10. Februar 2012

Bibliografische Information der Deutschen Nationalbibliothek

Die Deutsche Nationalbibliothek verzeichnet diese Publikation in der
Deutschen Nationalbibliografie; detaillierte bibliografische Daten sind
im Internet über http://dnb.d-nb.de abrufbar.

ISBN 978-3-8325-3138-6

Logos Verlag Berlin GmbH
Comeniushof, Gubener Str. 47,
10243 Berlin
Tel.: +49 (0)30 42 85 10 90
Fax: +49 (0)30 42 85 10 92
INTERNET: http://www.logos-verlag.de

Meinen Eltern

Contents

List of abbreviations

ADC	Analog digital converter
ARES	Advanced Rheometric Expansion System
Ca	Capillary number
CP	Cone plate geometry
ΔG_f	Gibbs free energy of flocculation
ΔH_f	Enthalpy of flocculation
ΔS_f	Entropy of flocculation
$\Delta \nu$	Spectral resolution
dbcp	Diblock copolymer
DC	direct current
D_e	Deborah number
DHR2	Discovery Hybrid Rheometer 2
DLVO	Derjaguin Landau Verwey Overbeek
DSC	Differential scanning calorimetry
$\underline{\underline{D}}$	Deformation tensor
ϵ	Separation parameter
E	Nonlinear mechanical master number
E^0	Nonlinear mechanical master number
E^0_λ	Viscosity ratio dependent nonlinear mechanical master number
E_A	Attractive energy

List of abbreviations

E_a	Energy barrier
E_d	Dissipated energy
E_R	Repulsive energy
E_t	Total energy
η	Viscosity
η_d	Matrix of dispersed phase
η_m	Matrix viscosity
F	Force
FFT	Fast Fourier Transformation
FT	Fourier Transformation
FT-IR	Fourier Transformation Infrared
$F_{12}^{(\dot\gamma)}$	Schematic model
Γ	Surface/interfacial tension
Γ_E	Interfacial tension determined by E
Γ_{MCT}	Decay rate
Γ_{meas}	Surface/interfacial tension determined with the Wilhelmy plate and ARES G2
Γ_{Pal}	Interfacial tension determined by Palierne model
Γ_{PD}	Interfacial tension determined by pendant drop measurement
Γ_{ref}	Reference surface/interfacial tensions
$\dot\gamma$	Shear rate
γ_{yield}	Yield strain
γ_0	Strain amplitude
γ_c	Critical strain amplitude for break up of neighbor cages
G	Shear modulus
G'	Elastic modulus
G''	Viscous modulus
HMCTS	Hexamethylcyclotrisiloxane
θ	Contact angle
Θ	Scattering angle
$I_{n/1}$	Relative intensity of higher harmonic n
$\underline{\underline{I}}$	Second rank unit tensor
I and II	First and second scalar invariant

I_n	Absolute intensity of higher harmonic n
k_B	Boltzmann constant
λ	Viscosity ratio
LAOS	Large amplitude oscillatory shear
μ	Mean value
M	torque
MCT	Mode Coupling theory
Miglyol 812	Capric/Caprylic Triglyceride
$m(t)$	Memory function
M_e	Entanglement molecular weight
M_n	Number average molecular weight
M_p	Peak molecular weight
M_w	Weight average molecular weight
ν_{Ny}	Nyquist frequency
NMR	Nuclear magnetic resonance
$\underline{\mathbf{n}}$	Unit vector normal to the interface
$\underline{\nabla}\,\mathbf{v}$	Undisturbed velocity gradient
o/w-emulsion	Oil-in-water-emulsion
p	Pressure
PDI	Polydispersity defined by $\frac{\langle R \rangle_{43}}{\langle R \rangle_{10}}$
PDI_{53}	Polydispersity defined by $\frac{\langle R \rangle_{54}}{\langle R \rangle_{43}}$
PDI_M	Polydispersity of the polymers defined by $\frac{M_w}{M_n}$
PDMS	Polydimethylsiloxane
PEG	Polyethylene glycol
PGPR	Polyglycerinpolyricinoleate
PI	Polyisoprene
PIB	Polyisobutylene
PP	Plate plate geometry
PNIPAAM	poly(N-isopropylacrylamide)
P_n	Polymerization degree
Pe	Dressed Peclet number, Weissenberg number
Pe_0	Peclet number

v_1, v_2	Static structure factors
$\underline{\mathbf{q}}$	scattering vector
$^{5/3}Q_0$	Intrinsic nonlinear ratio
$^{3}Q_0$, Q_0	Intrinsic nonlinear parameter
ρ	Density
ρ_n	Absolute phase of the nth harmonic
R	Radius
RI	Refractive index
$\langle R \rangle_{43}$	Volume average radius
R_H	Hydrodynamic volume
$S(\underline{\mathbf{q}})$	Wave vector dependent structure factor
S/N	Signal to noise ratio
σ	shear stress
σ_d	Standard deviation
σ_i	interfacial shear stress
σ_{yield}	Yield stress
$\underline{\underline{\mathbf{S}}}$	Droplet shape tensor
sr	Sampling rate
τ_α	Structural relaxation time
τ_β	Interfacial relaxation time
τ_{relax}	Characteristic relaxation time
T	Temperature
T_{VF}	Vogel-Fulcher temperature
t_{aq}	Acquisition time
T_b	Boiling temperature
t_{dw}	Dwell time
t_f	Flow time
T_g	Glass transition temperature
T_m	Melting temperature
$\underline{\mathbf{u}}$	Velocity at the interface
U	Perimeter
W_{min}	Minimum work

SAOS	Small amplitude oscillatory shear
SAXS	Small angle X-ray scattering
SEC	Size exclusion chromatography
THF	Tetrahydrofuran
UV	Ultraviolet
Φ_{Vol}	volume fraction
$\Phi(t)$	Density correlator
ϕ_n	Relative phase of higher harmonics
χ_{FH}	Flory-Huggins interaction parameter
$\frac{1}{\chi_D}$	Debye length
$\Psi(R)$	Droplet size distribution
Ω	Angular frequency
$\underline{\underline{\Omega}}$	Vorticity tensor
$\omega_1/2\pi$	Excitation frequency
w/o-emulsion	Water-in-oil-emulsion

Zusammenfassung

Dispersionen sind ein heterogenes Gemisch aus zwei Stoffen, die sich nicht miteinander mischen. In der vorliegenden Arbeit werden sowohl Suspensionen bestehend aus einer festen dispersen Phase in einem flüssigen Dispersionsmedium, Emulsionen bestehen aus zwei nicht mischbaren Flüssigkeiten als auch Schäume, gebildet aus einer Gasphase dispergiert in einer Flüssigkeit, untersucht. Als Hauptcharakterisierungsmethode dient die Fourier Transformations-Rheologie (FT-Rheologie), welches eine rheologische Methode im nichtlinearen Bereich darstellt.

Die rheologische Untersuchung der Emulsionen sowohl mittels Simulationen basierend auf dem elliptischen Modell von Maffettone und Minale in Kombination mit der Batchelor Theorie als auch der gleichzeitigen experimentellen Charakterisierung resultiert in der Erzeugung einer nichtlinearen mechanischen Masterkurve E für Emulsionen. Diese Masterkurve dient zur Bestimmung der Emulsionseigenschaften wie Tropfengröße, Tropfengrößenverteilung und Grenzflächenspannung.

Die mechanischen Eigenschaften der Suspensionen, bestehend aus einer wässrigen Dispersion thermosensitiver Kern-Schale-Partikel, unter oszillatorischer Scherung können mittels der Modenkopplungstheorie (MCT) beschrieben werden. In dieser Arbeit wurden die theoretischen Vorhersagen der MCT zum ersten Mal experimentell gemessen. Das heißt die Theorie konnte zum ersten Mal bestätigt werden als auch auf ihre Schwachstellen hingewiesen werden.

Auch bei den Schäumen wurden erstmals nichtlineare rheologische Experimente durchgeführt. Anhand von Schäumen verschiedener Biersorten wie Guinness, Köstritzer, Kilkenny, Kölsch, Rothaus Pils und Erdinger Weißbier konnte gezeigt werden, dass die sehr sensitive Methode der FT-Rheologie auch Systeme messen und differenzieren kann, die hauptsächlich aus Luft und einem Ethanol-Wasser-Gemisch bestehen. Zur Messung solcher Systeme wurde die Sensitivität der Messmethode zunächst optimiert. Damit konnte ein Signal-zu-Rausch-Verhältnis von 10^7 erreicht werden, was bezüglich des heutigen Entwicklungsstands der Rheometer als mechanische Charakterisierungsmethode einem Maximum entspricht.

Die vorliegende Arbeit umfasst die neuartige Charakterisierung verschiedener Dispersionen basierend auf der FT-Rheologie. Zusammenfassend kann gesagt werden, dass die FT-Rheologie eine vielversprechende Untersuchungsmethode für weiche Materialien ist, da ihre Sensitivität genutzt werden kann, um die verschiedenen mechanischen Eigenschaften aufzulösen wie z.b. im Fall der Schäume verschiedener Biersorten ein unterschiedliches Verhalten gemessen werden kann. Außerdem ist es möglich Aussagen über Tropfengrößen und Grenzflächenspannungen von Emulsionen zu erhalten und die Ergebnisse der Modenkopplungstheorie für Suspensionen zu validieren und nötige Verbesserungen aufzudecken.

1. Introduction

Dispersions consist of finely subdivided systems of a dispersed phase in a dispersion medium. The state of aggregate of the dispersed phase classifies different types of dispersions. Clear distinctions are made between the phase and the size of the dispersed medium. In this work, the dispersion medium is a liquid and, therefore, other media will not be discussed. A dispersion containing distributed solid particles is called a sol or suspension, while a liquid in a liquid matrix is called an emulsion and is investigated in Chapters 7. With a dispersed polymer phase, the dispersion is then called a polymer colloid and is explored in Chapter 6. The last type of dispersion consists of a gaseous dispersed phase in a liquid and is called a foam and is investigated in Chapter 9.

Dispersions can be found in many areas of products and build an important class of materials. Figure 1.1 shows several examples of suspensions, emulsions and foams as found in daily life. The processing and application conditions of most of these dispersions cause a change of the elastic and viscous properties of the material. In terms of rheology this applied load is within the so called nonlinear regime [Larson 99, Dealy 06, Mezger 06]. Examples are extrusion of mayonnaise while it is being packaged and shearing of either paint as it is applied to a wall or of cream as it is rubbed into human skin. Thus, nonlinear rheological investigations of model dispersions are important to better understand the behavior of the corresponding, very complex real-life systems.

Figure 1.1: Pictures of (a) suspension (paints), (b) emulsions (cream, mayonnaise or milk) and (c) foams (cappuccino, bubble bath or beer foam) from daily life.

1.1 General aspects about dispersions and their mechanical characterization

Dispersions are an integral of colloidal science, governing the properties of emulsions, suspensions and foams [Dörfler 02]. The origins of colloidal science can be traced to the middle of the 19th century and were pioneered by the scientists T. Graham, J. Tyndall, and M. Faraday. Yet it was only in the 20th century that colloidal chemistry was established. Graham (1805-1869) investigated the absorption and diffusion of gases through liquids as well as the permeation of colloids through membranes. His results showed that small ions can pass through membranes, but larger molecules or aggregates could not. This observation guided him to introduce a new notation for such macromolecules: 'colloids'. This term originates from the Greek word 'kolloi' and means to glue. Faraday (1791-1867) and his student Tyndall (1820-1893) discovered the light diffraction phenomena in colloidal dispersions and the change of polarisation in diffracted light. Both assumed that this was a typical characteristic of colloids. In the 20th century, this result was the fundamental historical work leading to the development of static and dynamic light scattering. Interfacial and colloidal phenomena were already used in ancient Egypt and in the Middle Ages as inks (colloidal pigment dispersion), detergents (surfactant), colored glasses (solid dispersion within a glass matrix of inorganic pigments), cosmetics (emulsion) and ointments (gel) [Brummer 06].

The numerical value for the diameter of the dispersed phase further subdivides the dispersions into different categories as visualized in Figure 1.2. The foams and dilute model emulsions investigated in this thesis are all macro-emulsions with a dispersed phase diameter larger than 5 µm. However, the commercial emulsions studied here are instead classified

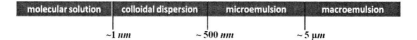

Figure 1.2: Classification of emulsions in dependence of the diameter of their diluted particles.

as micro-emulsions with droplet diameters smaller than 5 μm and typically larger than 500 nm. The polymer colloid has a diameter greater than 1 nm, [Crassous 08a], and is, therefore, also categorized as a colloidal dispersion. It should be noted, that the schematic scaling in Figure 1.2 is only a manmade guideline for classification and the transitions are, in reality, smooth.

Rheology as science is the study of deformation and flow of materials [Macosko 94]. Classical rheological measurements are applied either in the linear regime, where the excitation is small and on a time scale below the relaxation time of the investigated material. Increasing the frequency, shear rate or deformation amplitude of shear on the material, the nonlinear regime might be approached. Rheology is applied in a wide field of application and includes different types of experiments. Elongational flow is exposed in extensional rheometry, pressure driven flow in capillary rheometry and shear flow in rotational rheometry [Macosko 94, Mezger 06, Nakayama 99]. Within this thesis mainly rotational rheometry is applied and specifically with oscillatory measurements in the nonlinear regime. Oscillatory measurements are defined by the applied excitation frequency $\omega_1/2\pi$ and the strain deformation γ_0 and are extended into the nonlinear regime with Fourier Transform rheological experiments, substantially improved by Wilhelm et al. [Wilhelm 98, Wilhelm 02, Hyun 11].

The title of this dissertation is 'Fourier Transform Rheology on Dispersions based on Newtonian Fluids', which describes quite well the investigated field of research. The types of dispersions investigated have varied from hard spheres to soft deformable droplets and, finally, to gaseous bubbles as the dispersed phase, where FT-Rheology was applied to all these dispersions as the major characterization technique. A special feature of this work is that the experimental results were supplemented by simulations using constitutive equations, which resulted in contributions that extended the state of the art.

1.2 State of the art for mechanical characterization of the investigated systems

This thesis investigates three types of dispersions: emulsions, polymer colloids and foams. The FT-rheological characterization of the polymer colloids and the emulsions are coupled with theoretical calculations. Simulations of the polymer colloid nonlinear behavior is based on the mode coupling theory assuming hard spheres. In contrast, simulations of the emulsions are based on elliptical models describing the deformation of the droplets. As large amplitude oscillatory shear (LAOS) investigations of foams are very new, constitutive equations could not be found to describe and predict the nonlinear oscillatory behavior. First, an introduction to the different research projects is given.

FT-Rheology

Fourier Transform Rheology (FT-Rheology or FTR) explores the nonlinear regime using time dependent, large amplitude oscillatory shear experiments, [Hyun 11, Wilhelm 02]. FT-Rheology is a useful tool because it converts the measured shear stress (torque) data in the time domain into a frequency dependent spectrum that can detect even very weak nonlinearities in the stress response. FT-Rheology has been applied to many types of complex fluids in the literature and can, for example, distinguish between linear and branched polymers [Neidhöfer 04, Schlatter 05, Hyun 07, Hyun 09] and infer the droplet morphology in polymer blends [Grosso 07, Carotenuto 08]. The possibility to use FT-Rheology for the characterization of dispersions of colloidal particles is shown in an early study of Kallus et al. [Kallus 01] where the standard nonlinear steady shear experiments could not distinguish between two different aqueous polymeric dispersions. However, differences between these samples were clearly seen in the nonlinear oscillatory experiments.

Emulsions

The main part of this thesis concerns the nonlinear mechanical investigations of emulsions. Emulsions (and blends in general) are a class of materials where two or more fluids or viscous constituents are blended together to create a new 'composite' material. Blends can be roughly divided into three categories: miscible, partially miscible and immiscible blends. Immiscible blends are by far the most common group and are ubiquitous in the polymer and food processing, pharmacology and cosmetic industries [Brummer 06, Clemens 04, Derkach 09, Larson 99, Tucker 02]. For dilute emulsions, the typical microstructure of immiscible components at rest consists of spherical droplets immersed in a continuous matrix. The size and size distribution of these globular domains strongly affect both the processing and the mechanical properties of the final products. Therefore, significant efforts were made in the last decades to quantitatively measure and control the morphology of blends using non-invasive measurement techniques such as direct optical microscopy, light scattering and rheological methods, etc. [Fahrländer 99, Palierne 90, Palierne 91, Rusu 99]. For rheological experiments, the interest has been primarily focused on establishing protocols to determine the morphological properties like droplet size and droplet size distribution of emulsions, [Tucker 02, Rallison 84, Chesters 91, Stone 89, Ottino 99]. It is important to note that these experiments typically used either dynamic Small Amplitude Oscillatory Shear (SAOS) flow or steady shear flow and analyzed the mechanical linear or steady state response, e.g. $G'(\omega)$ (elastic modulus), $G''(\omega)$ (viscous modulus) or $\eta(\dot{\gamma})$ (shear rate dependent viscosity) [Mezger 06].

Most rheological models of dilute emulsions are based on ellipsoidal deformation models where the morphology of the included phase is assumed to be globular at rest and a single droplet is modeled as an ellipsoid under deformation. Figure 1.3 shows such a deformation of an isolated spherical droplet into an ellipsoid under LAOS. When dealing with neutrally-buoyant Newtonian droplets dispersed in a Newtonian fluid at low Reynolds numbers (where the Reynolds number is equal to the ratio of inertial forces to viscous forces, [Nakayama 99]), the relevant physical parameters under periodic shear flow conditions include the external fluid viscosity η_m, the droplet viscosity η_d, the interfacial tension Γ, the radius of the undistorted spherical droplet R and the maximum shear rate of the macroscopic flow $\dot{\gamma}_{max} = \gamma_0 \omega_1$, where $\omega_1/2\pi$ is the excitation frequency and γ_0 is the strain amplitude. Two dimensionless numbers are defined using these quantities, the capillary number $Ca = \frac{\eta_m \dot{\gamma} R}{\Gamma}$ and the viscosity ratio $\lambda = \frac{\eta_d}{\eta_m}$ [Reinheimer 11b]. The capillary number describes the relative importance of viscous to interfacial forces. The viscosity ratio is the ratio of the viscosity of the dispersed phase over the viscosity of the matrix. Both directly affect the degree of droplet deformation under shear as the interfacial tension resists the applied shear stress to preserve the original droplet shape.

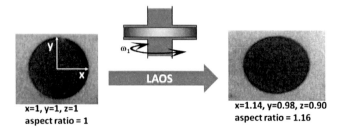

Figure 1.3: 2-D Image of a droplet deformation under large amplitude oscillatory shear (LAOS) with restriction of volume conservation. Comparison of the volume of the sphere ($V = \frac{4}{3}\pi x^3$ with
$x = y = z$) with the volume of the ellipsoid ($V = \frac{4}{3}\pi xyz$ with x, y, z =semiaxes of the ellipsoid)
leads to $z = 0.90$ of the deformed droplet [Bronstein 08]. The axes of the undistorted droplet
are set to unity. The measurement was made in a transmission light microscope equipped with
a shear cell. The photo was taken with a camera equipped with a macrolens. The microscope
had an objective of 20 fold magnification and an ocular of 10 fold magnification. The here
sheared emulsion consists of a dispersed phase of 50 wt% polyethylene glycol solution in water ($M_w = 20\,000$ g/mol purchased by Carl Roth GmbH + Co. KG) in the continuous phase
miglyol 812 (see Chapter 5). Fluorescein sodium salt was added for a better contrast. The
excitation frequency was 1 Hz and the deformation amplitude $\gamma_0 = 100$. The photo was taken
at maximum deflection of the oscillatory shear.

The existence of a clear connection between the morphology of the emulsions (i.e. the
shape of the droplets and their orientation in flow) and their rheological behavior has been
previously established [Maffettone 98, Bousmina 00, Jansseune 00, Jackson 03, Guido 04,
Yu 03, Almusallam 04]. Within the simulations the three semiaxes of an ellipsoid have
different values, thus the rotational symmetry of the droplets is lost. Under deformation
the ellipsoids possess two mirror planes of symmetry.

Numerical simulations, based on the boundary element technique, give very accurate an-
swers, but are, unfortunately, very time consuming and require significant computational
resources [Kennedy 94, Zinchenko 97]. Several analytical models for droplet deformation
have been proposed in the literature as an adequate and simple way to describe droplet dynamics under generic flow conditions [Maffettone 98, Yu 03, Jackson 03, Almusallam 04].
One particularly effective phenomenological model introduced by Maffettone and Minale
[Maffettone 98], hereafter referred to as the MM model, was proven to give quantitative
agreement with experimental results up to Ca values slightly larger than unity [Guido 04].

The MM model is also a valuable tool to infer emulsion properties such as the volume average droplet radius, the droplet distribution and interfacial tension under these conditions as shown within this thesis and in Jansseune et al. [Jansseune 00].

For the situation where SAOS is applied, Palierne proposed a mathematical solution for the deformation and orientation of liquid droplets in flow [Palierne 90, Palierne 91, Graebling 93a, Graebling 93b, Lacroix 96]. In these studies, the dependence of the viscoelastic properties of emulsions on parameters such as matrix viscosity, interfacial tension, the radius of the system and its oscillation frequency were investigated in detail. The Palierne model led to a robust procedure that could relate the linear viscoelastic properties of polymer blends to their chemical-physical properties. For example, the model could estimate the volume average droplet radius of a blend [Das 05] or, alternatively, its interfacial tension [Huitric 49, Guschl 03, Vincze-Minya 07]. The Palierne model has been subsequently extended to estimate a finite volume fraction [Palierne 90, Palierne 91], to calculate a droplet size distribution instead of an averaged mean value [Friedrich 95] and to include the effect of surfactants on the interfacial tension and the droplet deformation [Jacobs 99]. The impact of the surfactant concentration at the interface on the resulting droplet deformation is of great interest, for example in food science applications [Windhab 05]. Within this thesis a new protocol to infer the average droplet size and information about the width of the distribution will be based on FT-rheological protocols. It will be shown, that this newly invented procedure, based on LAOS measurements, has the advantage of unlimited polydispersity of the droplet size distribution, whereas the linear analysis is restricted to narrow distributions, see Chapter 7.

The main aspect of this thesis is the extension of the characterization of emulsions into the nonlinear regime using Large Amplitude Oscillatory Shear flow. The intensities and phases of the higher harmonics in the amplitude spectrum provide valuable information on the sample being evaluated. These higher harmonics are defined as the intensities appearing at odd multiples of the excitation frequency $\omega_1/2\pi$. The purpose of this work is to establish a new nonlinear mechanical coefficient, E, for emulsions where E is proportional to the ratio of the 5th harmonic intensity to the 3rd harmonic intensity scaled by the square of the capillary number. This proposed coefficient E quantifies the nonlinearities arising from the interfacial stress contribution, which is, after all, mainly responsible for the non-Newtonian behavior of the emulsions under investigation.

Colloidal dispersions

As a further topic of this thesis, FT-Rheology is applied to polymer colloids. Polymer colloids exist in fluidic phases, glassy phases and as crystals. The main parameter for the state of aggregate is the volume fraction of the dispersed phase as shown in Figure 1.4. This work is based on a cooperation of three parties consisting of the preparative

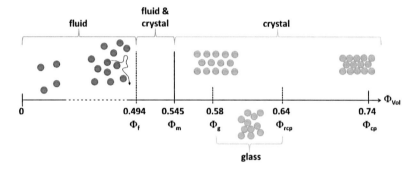

Figure 1.4: Phase diagram of the dispersed phase as a function of the volume fraction Φ_{Vol} for a dispersion of hard spheres according to [Poon 95]. The different specified volume fractions Φ_{Vol} are Φ_f indicating the fluid transition, Φ_m for the melting transition, Φ_g for the glass transition, Φ_{rcp} for random close packing and Φ_{cp} for close packing.

contribution by Professor Ballauff (Helmholtz-Zentrum Berlin, Germany), the theoretical part calculated by Professor Fuchs (University Konstanz, Germany) and Professor Brader (University Fribourg, Switzerland) and the nonlinear rheological investigation done as part of this thesis in the group of Professor Wilhelm. The cooperation focused on the investigation of colloidal dispersions close to or in the glassy region. Thermosensitive core-shell particles dispersed in water were used to investigate the nonlinear rheological behavior of dense colloidal suspensions [Siebenbürger 09]. As the polymer phase in colloids is normally in a glassy state below the glass temperature T_g, they are often called colloidal suspension. The shell size of the dispersed particles changes with temperature. Thus, the volume fraction can be adjusted from close to until the point of dynamical arrest, by varying the temperature [Brader 10]. The glassy region is induced by the cage effect (where particles are closely surrounded by adjacent particles), which hinders the particles from free

Brownian motion. In linear rheology, these systems exhibit strong elastic behavior, whereas under large amplitude oscillatory shear, plastic flow is observed. In practical applications, large deformations are of special interest as explained before. Several theories present a theoretical treatment of the nonlinear regime under oscillatory shear. Phenomenological models are based on the principles of continuum rheology as presented in the following references [Larson 88, Ewoldt 08, Klein 07, Cho 05, Bird 87].

The mode coupling type theories [Götze 92] and references therein [Brader 07, Brader 08, Fuchs 02, Fuchs 09] capture the slow structural relaxation process, which leads to a dynamical arrest in strongly coupled systems. The microscopic constitutive equation as proposed in [Brader 08] is generally applicable to theory, but is difficult to handle in numerical simulations of both quiescent systems and systems under steady shear flow. However, the schematic $F_{12}^{(\dot{\gamma})}$ model, derived in [Brader 09], captures all the physics of a fully microscopic theory. Therefore, the derivation of the mode coupling theory (MCT) and its development to the schematic $F_{12}^{(\dot{\gamma})}$ model for predictions of large amplitude oscillatory shear behavior will be one of the topics of this thesis. For the first time a theory based on the microscopic MCT was able to predict the experimental behavior of a polymer colloid close to and beyond the glass transition under LAOS. Experiments were applied to a thermosensitive colloidal suspension in the nonlinear rheological region.

Foams

In the last part of this thesis, closed foams are explored using FT-rheological experiments. For the first time it was shown that FT-Rheology was able to resolve differences between various brands of beers. Foam rheology is of high interest, but investigations are typically made under steady shear or within the linear oscillatory regime [Kraynik 88, Princen 82, Princen 85, Tcholakova 08, Denkov 09] and references cited there. The problem for LAOS measurements is the detection of the torque. Foams with a high volume fraction of dispersed phase consist mainly of gas and, therefore, generate a LAOS response with generally a low torque of dimension µNm. Therefore, the LAOS measurements have to be performed with a very sensitive rheometer. Within this thesis the most advanced rheometer ARES G2 from TA Instruments was used, which possess the highest sensitivity of status quo strain controlled rheometers, see also Chapter 3, and is able to measure torques down to 50 nNm. The requirement of a high torque resolution limited previous investigation of nonlinear oscillatory shear experiments and makes the here proposed experimental results unique and pioneering.

Steady shear deformations are generally described using the three parameter Herschel-Bulkley model:

$$\sigma = \sigma_{\text{yield}} + \sigma(\dot{\gamma}) = \sigma_0 + k \cdot \dot{\gamma}^n \tag{1.1}$$

where k is the consistency parameter, n the power law index, $\dot{\gamma}$ the applied shear rate, σ the total shear stress and σ_{yield} the yield stress, [Princen 01, Tcholakova 08]. The flow properties of foams are determined by strong elastic forces. Bulk rheology is simulated with geometrical models considering periodic hexagonal structures for highly concentrated systems as shown in Figure 1.5 [Princen 82]. The model assumes as a first approximation monodisperse regular structures. The periodic structure of a monodisperse foam with

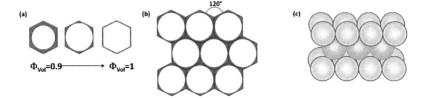

Figure 1.5: a) Increasing volume fraction of the gaseous dispersed phase leads to an increasingly hexagonal structure. b) Two dimensional periodic structure assumed in highly concentrated monodisperse foams as proposed by [Princen 82]. c) Three dimensional face centered cubic structure for highly concentrated monodisperse foam with twelve neighbors as it is assumed in theoretical models by for example Denkov et al. [Denkov 09] and Tcholakova et al. [Tcholakova 08].

defined angles of $120\,^{\circ}\text{C}$ in a two dimensional model results of balanced tensions between the bubbles and the confining walls. Deformation of the droplets are resisted by preserving the static angle. When the yield stress is exceeded, the static balance is lost and the bubbles can slide with respect to each other and generate viscous and dissipative behavior. From this description the linear oscillatory behavior at small strain amplitudes can be predicted where the elasticity is preserved during oscillatory deformation. The bubble structure is merely deformed and restoring effects dominate the shear stress. A three dimensional model considering a face centered cubic structure with twelve neighbors is proposed by Tcholakova et al. [Tcholakova 08], see Figure 1.5c. It is believed, that the LAOS experiments performed within this thesis are some of the first experiments in the non-static region. The FT-rheological experiments presented in this thesis were applied on beer foam and exhibit interesting results, thereby opening a new research area for further investigations. These experiments were difficult because of the complexity of the foams used in commercial products due to their polydisperse structure as well as their unstable structure.

1.3 Structure of this thesis

The introduction to FT-Rheology in Chapter 2 presents the reasons why it was necessary to have an in depth understanding of the role of the signal to noise ratio (S/N) in mechanical measurements. Therefore, an extra chapter (Chapter 3) was added, to discuss strategies from improving S/N. Over the course of this thesis, different generations of rheometers were used to measure FT-Rheology. It was found that the development of the rheometers influenced the experimental procedure as well as the lower resolution limit. It will be shown that these improvements in the resolution limit resulted in both a higher S/N, Section 3.2, and in an increase in the maximum detectable higher harmonic in the frequency spectrum, Section 3.4.

Several theoretical aspects concerning dispersion stabilization are discussed in Chapter 4. Stability is an important parameter in the characterization of dispersions because reproducible measurements can only be achieved with stable systems, whereby stability in the dimension of minutes is necessary. Stability is governed by attractive forces and their influence on aggregation, flocculation and coagulation, see Figure 4.1 [Piirma 92]. Therefore, the dispersion stability is also influenced by the surface and interfacial tensions. In addition, the concept of interfacial tension, its definition, its derivation and its occurence in daily life phenomena are discussed in Chapter 4. Another topic of interest is steric stabilization, which explains the way block copolymers stabilize emulsions. Related to this topic, additional surface and interfacial tension measurements were conducted on defined systems with a new setup consisting of a combination of the rotational rheometer ARES G2 and a Wilhelmy plate of platin-iridium alloy, see Chapter 5.

After the introduction to the theory of FT-Rheology and dispersion stability, the optimization of the rheological experiments led to the application of LAOS to different dispersion systems. In Chapter 6, FT-Rheology is applied to colloidal suspensions. The theoretically calculated nonlinear response is then compared with experimental data. The simulations from the schematic $F_{12}^{\dot{\gamma}}$ model are based on the microscopic MCT approach. It will be shown that simulations based on the microscopic theory have a predictive capacity for the macroscopic properties. Afterwards, the dispersed phase aggregate state is changed from a solid to a liquid in the chapter of emulsion rheology. This chapter introduces first the constitutive models to simulate the nonlinear rheological behavior of emulsions (Section 7.1). Sections 7.3 to 7.6 summarize the definition and development of E as a function of the capillary number Ca and the ratio of the viscosities of the two emulsion components as based on the numerical simulations. The influence of polydispersity is also considered. Experimental results are presented in Section 7.7 for two different dilute polymer blends and commercial water-in-oil emulsions (w/o-emulsions). With the addition of diblock copolymer, which acts as a compatibilizer for the two immiscible fluids, the interfacial tension of the dilute polymer blends could be varied. For the work reported here, a commercial diblock copolymer PIB-b-PDMS was purchased and an anionically polymerized PI-b-PDMS was synthesized (Section 8.4) and both were used to influence the interfacial tension of a polymer blend consisting of polydimethylsiloxane (PDMS) in polyisobutylene (PIB) or polyisoprene (PI), respectively, Section 7.7.1. The use of anionic synthesis resulted in a deblock copolymer with a defined length and composition. The advantages of anionic synthesis to generate defined structure as well as relevant theoretical aspects are given in Sections 8.1 and 8.2.

To complete this thesis, foams were investigated as a last class of dispersion Chapter 9. While this thesis, the FT-rheological behavior of beer foam was explored in first experiments, it could be shown that FT-Rheology is a useful tool to distinguish between different brands of beer. However, there is still a fundamental lack of understanding of the underlying mechanisms and these first results are presented together with several open questions that should stimulate future research.

2. Fourier Transform Rheology

2.1 Fourier Transformation in rheological analysis

This theory part describes the most important aspects of the Fourier transformation (FT) needed for executing FT-Rheology and is rephrased from the literature [Wilhelm 02]. If the time dependent signal, for example torque, is a continuous and continually integrable function, which contains periodic contributions, the FT converts these periodic contributions into the frequency domain either into their magnitude and phases or their real and imaginary part. It is assumed that the sample constitution does not change significantly within the time scale of the deformation period T_p of an FT-rheological experiment. In these cases more complicated transformations, like wavelet transformation [Fearn 99, Honerkamp 94], would be necessary and are therefore omitted. Fourier Transform Rheology applies a half sided, discrete, complex, magnitude Fourier transformation of the shear stress data. The mathematical definition of Fourier transformation of any real or complex time signal, $s(t)$, or frequency dependent spectrum $S(\omega)$ is given in Equation (2.1):

$$S(\omega) = \int_{-\infty}^{\infty} s(t)e^{-i\omega t}dt$$
$$s(t) = \frac{1}{2\pi} \int_{-\infty}^{\infty} S(\omega)e^{+i\omega t}d\omega \ . \tag{2.1}$$

The prefactors may differ depending on the used convention. Generally the FT is an invertable, linear, complex transformation over the infinte interval from $-\infty$ to $+\infty$. The basic mathematical idea behind Equation (2.1) is as follows: a set of functions, e.g. polynomial, Legendre-, Laguerre-, Hermite-polynomial or as used for the FT harmonic functions can span, in close similarity to vectors, a space where the different functions act basically as orthogonal vectors. The class of oscillating functions is orthogonal with respect to all different frequencies. The space therefore has an innumerable infinite dimension, when the infinte interval is considered [Wilhelm 02]. The time dependent signal $s(t)$ is sorted by the FT with respect to the inherent frequencies $\omega/2\pi$ with their corresponding amplitudes and phases in a spectrum $S(\omega)$. Other important mathematical properties of the FT are described below. First of all its linearity, which means that any superposition of different signals in the time domain, for example $s(t)$ and $g(t)$, will also be a superposition in the frequency domain:

$$a \cdot s(t) + b \cdot g(t) \longleftrightarrow a \cdot S(\omega) + b \cdot G(\omega) \, . \tag{2.2}$$

The FT is inherently complex. Due to this fact, even a real time-domain data set $s(t)$ will result in a complex spectrum $S(\omega)$ with a real, Re, and imaginary part, Im, of the spectrum. Another possibility to present the complex spectrum is the magnitude $m(\omega)$ and phase $\phi(\omega)$ spectra, whereas $\tan(\phi(\omega)) = \frac{\text{Im}(S(\omega))}{\text{Re}(S(\omega))}$ and $m(\omega) = \sqrt{\text{Re}(S(\omega))^2 + \text{Im}(S(\omega))^2}$. Due to the Euler relation, $\exp(i\phi) = \cos(\phi) + i\sin(\phi)$, the FT Equation (2.1), can also be separated into a cosine- and sine transformation, also called Fourier cosine and Fourier sine transformation. The half sided FT implies an integration only along half the integral limits, specifically from $t = 0$ to $t = +\infty$. This is a common procedure for FT of experimental data. The last expression discrete FT will be explained in the following. Experimental data is normally not acquired continuously but discrete. They are taken point by point with a fixed increment t_{dw} called dwell time, or inverse sampling rate, sr. If N_p data points in time are measured, a total acquisition time is given by $t_{\text{aq}} = N_p \cdot t_{\text{dw}}$. The Fourier Transformation of this discrete data set (N points) is therefore called discrete FT and creates a spectrum with N_p complex points. The spectral width, respectively the maximal detectable frequency is called Nyquist frequency and determined by the sampling rate $\omega_{\text{max}}/2\pi = \nu_{\text{max}} = 1/(2t_{\text{dw}})$. The spectral resolution, defined as the frequency difference between two consecutive data points, is given by $\Delta\nu = 1/t_{\text{aq}}$. With the definition of the Nyquist frequency the desired maximum possible harmonic contribution should be estimated and the sampling rate accordingly adjusted [Wilhelm 02]. As an example an excitation with oscillation frequency of 1 Hz yields a higher harmonic contribution in the response spectrum up to the 25th overtone, if a minimum sampling rate of $50 \, \text{s}^{-1}$ is applied. Be aware that this sampling rate is meant to be after oversampling is applied. Oversampling reduces random noise and is used to increase the signal to noise ratio, Section 3.1, and is well explored in Dusschoten et al. [van Dusschoten 01, Hilliou 04, Skoog 92]. Like in every real spectrum, the peaks show a certain width and are not infinitely narrow. By increasing the acquisition time the observed line width decreases and the signal to noise ratio S/N increases since the integral is preserved. Here the S/N ratio can be defined as the ratio of the amplitude of the highest peak (signal S) divided by the standard deviation of the noise (noise N). The most intensive peak is generally at the shear excitation frequency, because the DC-component is ignored. The direct current (DC) component is the intensity at $0\,\mathrm{Hz}$ and reflects the offset of the signal in the time domain. To measure and quantify the noise level, a part of the spectrum without any peaks is chosen and the standard deviation calculated. For a sufficient S/N ratio each measurement is recorded about 5-50 cycles of the excitation frequency [Wilhelm 02]. This leads to a typical overall size of 1000 up to 10 000 data points N. By increasing t_{aq} the S/N ratio grows. The explanation is based on Equation (2.1). First the integral over the spectrum $S(\omega)$ with the time data $s(t)$ at $t = 0$ is compared:

$$s(0) = \frac{1}{2\pi} \int_{-\infty}^{\infty} S(\omega) \underbrace{e^{(+i\omega 0)}}_{1} d\omega = \frac{1}{2\pi} \int_{-\infty}^{\infty} S(\omega) d\omega \ . \tag{2.3}$$

As obvious from Equation (2.3), the value of the time dependent signal $s(t)$ at $t = 0$ cannot change as a function of the acquisition time t_{aq} and consequently the integral over the spectrum cannot change either. Therefore, an increasing t_{aq} increases the density of points in the spectrum, that means $\Delta\nu$ decreases. The integral is the width times the height and the width is reduced with $\frac{1}{t_{aq}}$ since all intensity is at one frequency data point. In the case of a forced oscillation under steady state conditions, the forced oscillation itself ideally lasts forever. Further details can be found in specific literature [Bracewell 86, Ramirez 95, Schmidt-Rohr 94, Claridge 99].

In experiments the S/N ratio can be increased by two possibilities, which reduce the statistical noise. First, like in other FT-techniques, multiple spectra can be averaged, where $S/N \propto \sqrt{n_c}$ in the case of stochastical noise with n_c being the number of cycles. This method leads to an increased precision of the relative intensity, but the phase angle information of the harmonic contributions is lost. The reason is, that in FT-rheology the response does not start for all frequencies at a time $t = 0$ with a phase $\phi_n = 0$. But the periodicity of the signal implies that the relative phase of the harmonics is fixed. Therefore

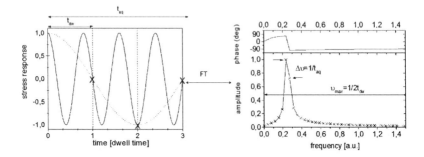

Figure 2.1: The Fourier Transformation of the discrete time signal with a dwell time t_{dw} leads to ambiguities with respect to different frequencies in time and frequency domain. The spectral width is the maximum frequency unambiguously determined by t_{dw}. The minimum resolution $\Delta\nu$ is limited by the acquisition time t_{aq} [Wilhelm 99].

the time data is consequently shifted with respect to the excitation and after FT the phase behavior can be analyzed. A second possibility is the already described oversampling of data points in Section 3.1 and in more detail in Dusschoten et al. [van Dusschoten 01]. In principle with both methods an unlimited S/N ratio for purely stochastic noise and disturbances can be achieved. The important concepts of dwell time, acquisition time, spectral width and spectral resolution are visualized in Figure 2.1. In this Figure 1000 time data points for 25 oscillations have been calculated prior to a Fast Fourier Transformation (FFT). In most experiments the time data is not measured continuously but discretely after fixed time steps and is then digitized. These analog time data are digitized every dwell time t_{dw} with a k-Bit analog-to-digital converter (ADC). This k-bit ADC has 2^{k-1} discrete values to discriminate the intensity of one point in time. Higher values of k lower the minimum detectable intensity of weak signals [Skoog 92, Homans 89, Claridge 99]. Thus a low bit ADC can be the limiting factor for measurements where a high S/N ratio is desired. The spectrum is calculated via FFT, which is a very common and particularly fast algorithm for the discrete FT. In FFT the calculation time rises only as $N_p \log_2 N_p$ in contrast to N_p^2 in a brute force discrete FT approach. Another significant difference is the number of required data points. The most common FFT ('butterfly') algorithm requires $N_p = 2^{n_p}$ data points with n_p as an integer value larger than zero and not arbitrary numbers like in the discrete FT [Cooley 65, Higgins 76]. As a consequence this leads to fixed values for t_{aq} and thus also for the spectral resolution $\Delta\nu = 1/t_{aq}$. Accordingly the fundamental frequency $\omega_1/2\pi$ or the odd multiples at $n\omega_1/2\pi$ are rarely located exclusively at a single data point with the precise frequency of multiples of the fundamental excitation in the frequency spectrum. Considering an integer number of periods within the finite acquisition time, that means full oscillation cycles, the frequency spectra shows a minimized leakage effect [Giacomin 98, Kammeyer 02]. The application of FT-Rheology within this thesis requires an optimal signal to noise ratio to detect the nonlinear intensities, see also Chapter 3. Therefore, the applied FFT-algorithm should not use automatic zero-filling to generate artificially 2^{n_p} time data points prior to the transformation into the frequency domain [Bracewell 86, Ramirez 95]. Thus, the 'butterfly' algorithm cannot be applied within the FT-Rheology used here.

2.2 Rheology: FT of a time dependent stress signal

In most cases mechanical properties are tested with rheological experiments in the linear regime as a function of temperature and frequency. The complex response function is described in terms of its real part (G') and imaginary part (G''). Physical properties like relaxation times or phase transitions of the non-perturbated samples can be evaluated. The linear rheology is characterized by the measurement of the elastic and viscous moduli G' and G'', respectively, as a function of strain amplitude γ_0 or excitation frequency $\omega_1/2\pi$. The linear rheology is described in detail in several textbooks [Macosko 94, Larson 99, Dealy 06, Mezger 06] and is here not further discussed. Within this thesis the nonlinear rheology is the major characterization technique. In most industrial processing conditions the shear rate exceeds the inverse of the longest relaxation time, which results in a change of the sample. This means the strain amplitude dependent elastic modulus $G'(\gamma_0)$ varies with increasing γ_0 as well as the viscosity as a function of the applied shear rate $\dot{\gamma}$ changes. Oscillatory measurements with large strain amplitudes achieve structural changes of the investigated sample like the deformation of a spherical droplet to an ellipsoid in emulsions [Larson 99, Guido 04]. The nonlinearity under oscillatory shear is measured in odd higher harmonic contributions at $(2n+1)\omega_1/2\pi$ of the excitation frequency $\omega_1/2\pi$ in the frequency dependent shear stress spectrum under oscillatory conditions, see Figure 2.2b. The nonlinear regime is characterized by the changing of the viscosity as a function of the applied shear rate under steady shear conditions, and the generation of odd higher harmonic contributions at $(2n + 1)\omega_1/2\pi$ in the frequency dependent shear stress spectrum under oscillatory conditions.

In a simple rheological experiment a force F acts on two planar and parallel surfaces with the area A at a distance d, where the sample is located in between. If a constant velocity ν and stress σ is reached the viscosity η and the shear rate $\dot{\gamma} = \nu/d$ might be defined by Newton's equation [Larson 99]:

$$\frac{F}{A} = \sigma = \eta \cdot \frac{\nu}{d} = \eta \cdot \dot{\gamma} \, . \tag{2.4}$$

In the linear region a Newtonian fluid has a constant viscosity η, whereas in the nonlinear behavior the viscosity becomes a function of the applied shear rate and time $\eta = \eta(\dot{\gamma}, t)$. As a simplification only one dimensional scalar values are considered. With the assumption of steady shear also for periodic conditions the time dependence can be neglected. In general no dependence on the direction of the applied motion is expected and therefore $\eta = \eta(|\dot{\gamma}|) = \eta(+\dot{\gamma}) = \eta(-\dot{\gamma})$. The shear thinning behavior respectively the transition from Newtonian to power law behavior under steady shear conditions can mostly be described with the Carreau model: $\eta = \frac{\eta_0}{1+(\beta\dot{\gamma})^x}$ [Macosko 94]. Under oscillatory shear with small nonlinear effects and the assumption of a steady shear under the periodic conditions, a polynomial or Taylor expansion of the viscosity with respect to the shear rate can be done [Wilhelm 02, Wilhelm 99]

$$\eta = \eta_0 + a\dot{\gamma}^2 + b\dot{\gamma}^4 + \dots . \tag{2.5}$$

The expansion coefficients η_0, a and b are complex. For the case of oscillatory shear, the deformation γ is given by:

$$\gamma = \gamma_0 \cdot e^{i\omega_1 t} \tag{2.6}$$

and the absolute vlaue of the shear rate respectively by:

$$\dot{\gamma} = i\omega_1\gamma_0 \cdot e^{i\omega_1 t} \, . \tag{2.7}$$

A short description of Fourier analysis can be found above and detailed in [Bracewell 86, Ramirez 95, Butz 98]. The Fourier analysis gives the time dependence of $|\dot{\gamma}|$ and thus the contribution to the shear rate dependent viscosity in the nonlinear regime. Therefore, in the nonlinear regime odd higher harmonics are measured in the frequency spectrum of the shear stress, as it is shown in Figure 2.2b. The appearance of these overtones will

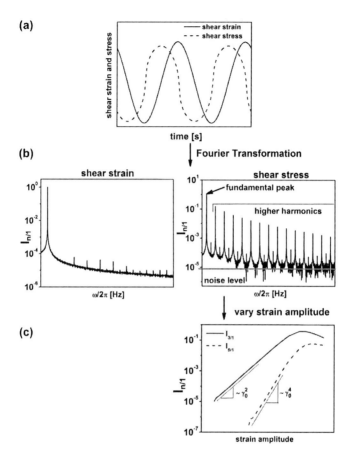

Figure 2.2: Schematic picture of the procedure of measuring and analyzing FT-Rheology. a) Measurement of the oscillatory shear strain and shear stress response in the time domain of the highly concentrated emulsion w/o-2 (Section 7.7.1) with an excitation of 1 Hz and $\gamma_0 = 10$. b) Normalized frequency spectra after the Fourier transformation of the shear strain and shear stress exhibit the fundamental peak at the excitation frequency $\omega_1/2\pi$. The odd higher harmonics $I_{n/1}$ with n being an odd multiple of one are measured only for the periodic nonlinear shear stress. The shear strain is the exciation and is assumed to be linear since only higher harmonics with $I_{n/1} < 10^{-4} = 0.01\%$ are measured. c) Visualization of the strain amplitude dependence of the first two higher harmonics $I_{3/1}$ and $I_{5/1}$ of simulated data where the simulation conditions were $\eta_m = 60\,\text{Pas}$, $\lambda = 3$, $\omega_1/2\pi = 0.1\,\text{Hz}$ and $\Gamma = 3\,\text{mN/m}$.

be deduced with the following substitution process. The viscosity related force, Equation(2.4), is rewritten by inserting Equation (2.7) into Equation (2.5) and completed with the substitution of this expression into Equation (2.4):

$$\sigma = (\eta_0 + a \; i^2 \omega_1^2 \gamma_0^2 e^{i2\omega_1 t} + b \; i^4 \omega_1^4 \gamma_0^4 e^{i4\omega_1 t} + \ldots) i\omega_1 \gamma_0 e^{i\omega_1 t} \ldots \qquad (2.8)$$

$$= \underbrace{\eta_0 \cdot i\omega_1 \gamma_0 e^{i\omega_1 t}}_{I_1} + \underbrace{a \cdot i^3 \omega_1^3 \gamma_0^3 e^{i3\omega_1 t}}_{I_3} + \underbrace{b \cdot i^5 \omega_1^5 \gamma_0^5 e^{i5\omega_1 t}}_{I_5} + \ldots$$

with $I_1 \propto \gamma_0^1 \omega_1^1$, $I_3 \propto \gamma_0^3 \omega_1^3$, $I_5 \propto \gamma_0^5 \omega_1^5$, etc..

Equation (2.8) expresses the mathematical transformation of the stress in the time domain into a frequency spectrum due to Fourier analysis. The time evolution leads to distinct signals at exactly the fundamental frequency $\omega_1/2\pi$ or odd higher harmonics $n\omega_1/2\pi$ in the spectrum. Each peak gives information about the magnitude, also referred as intensity I_n and the corresponding phase ϕ_n with respect to the phase of the fundamental peak at the excitation frequency (with n as the odd multiple of the excitation frequency) which are described in more detail in the following two sections.

2.2.1 Quantification of nonlinearity

In the early days of nonlinear experiments the nonlinear behavior of the shear stress is visualized in Lissajous curves. Lissajous diagrams represent the shear stress in the time domain $\sigma(t)$ as a function of the imposed deformation $\gamma(t)$, [Ewoldt 08, Hyun 03]. Purely viscous behavior is measured as a sphere if both, stress and strain, are normalized themselves and with respect to each other, whereas purely elastic behavior results in a line through origin. Linear Viscoelastic behavior appears as an ellipse with three symmetry elements, two mirror planes at the major and minor axis of the ellipse and symmetry about the origin. In the nonlinear region, the viscoelastic response loses its mirror symmetry, but shows a symmetry about the origin. This graphical illustration of nonlinear behavior gains only qualitative results Figure 2.3. Additionally the transition from linear to nonlinear behavior is not well characterized and difficult to measure. The visual detections of the deviation from ellipses might depend on the line thickness of the curves. Thus, the starting point where nonlinearity is revealed, depends on the illustration and the observer. It is obvious, that the quantification of the nonlinearity is improved substantially with the development of FT-Rheology, especially in the regime of low nonlinearity, see Figure 2.4.The Lissajous figure in Figure 2.4a shows a deviation from the ellipse with a third harmonic

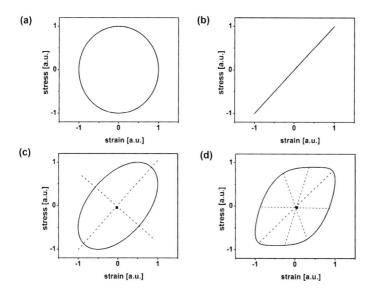

Figure 2.3: Schematic illustration of the Lissajous plots to the corresponding (a) visous, (b) elastic, (c) viscoelastic and (d) nonlinear rheological behavior. For the nonlinear rheological behavior 10 % third harmonic is assumed.

contribution of 4 % with the chosen visualization conditions. The nonlinear contribution can be amplified by plotting the shear stress as a function of the derivative of the strain, namely the strain rate, which is then also called viscous Lissajous plot especially applied for liquids, see Figure 2.4b [Tee 75, Giacomin 98]. In Figure 2.4c the minimum amount of third harmonic contribution was determined to be 2 % to achieve a visible loss of mirror symmetry.

In Figure 2.2b the noise level of a typical frequency spectra after Fourier Transformation of the shear stress in the time domain yields a noise level of 10^{-5}. Thus, a higher harmonic contribution with intensities of $I_{n/1} > 0.01\%$ can be theoretically detected and clearly distinguished from the noise. This means an improvement in quantification of the non-linearity of a factor of 10^2 with respect to the graphical analysis via Lissajous diagrams.

Other FT-rheological analysis are proposed in literature like for example in [Klein 07,

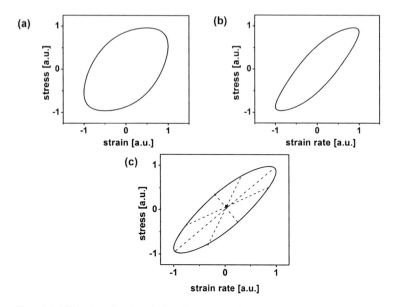

Figure 2.4: (a) Lissajous plot where the loss of symmetry with respect to the mirror planes becomes visible. The nonlinear contribution is 4 % third overtone $I_{3/1}$. (b) Shear stress as a function of the derivative of the strain, also called viscous Lissajous plot, shows clearly the nonlinearity of 4 % third harmonic contribution. (c) In the viscous Lissajous plot $\sigma(\dot{\gamma})$, [Tee 75, Giacomin 93b, Giacomin 98], the nonlinear contribution can be lowered to 2 % relative third harmonic $I_{3/1}$ as minimum nonlinear contribution for a graphical detection with the here chosen visualization conditions.

Ewoldt 08]. Ewoldt et al. applied the idea of Cho et al. [Cho 05], which is based on orthogonal stress decomposition. The shear stress is divided into a purely visous and a purely elastic contribution. The use of nonlinear constitutive equations include models for viscous and elastic behavior as well as the viscoelastic Giesekus model [Larson 88]. In the application of Klein et al. the whole frequency spectra is considered in a superposition of different overtone spectra of typical nonlinear rheological effects, like strain hardening, strain softening and shear bands or wall slip. Thus a first attempt was made to correlate the nonlinearities with underlying physical phenomena via four characteristic functions described in [Klein 07].

Next a quantitative description under oscillatory shear shall be introduced. The quantification is achieved via the intensities I_n and phases ϕ_n, see Equation (2.18), as a result of the Fourier analyzed LAOS measurements. The phases are defined in the next paragraph.

Intensities

The mathematical complex notation Equation (2.8) is rewritten to the real sinusoidal motion:

$$\sigma(t) = I_1 \cos(\omega_1 t + \rho_1) + I_3 \cos(3\omega_1 t + \rho_3) + \dots \; . \tag{2.9}$$

Within this thesis the definition of the intensity and phase which will be given in the following are based on [Wilhelm 99, Neidhöfer 03], whereas different interpretations can be found in [Klein 05, Giacomin 98].

The quantification of nonlinearity is achieved by the ratio of the nth harmonic to the 1st harmonic in the frequency spectrum to evolve an intense quantitiy for reproducible measurements:

$$\frac{I_n}{I_1} = \frac{I_{n\omega_1}}{I_{\omega_1}} = I_{n/1} \; . \tag{2.10}$$

The strain amplitude dependence of the higher harmonics at small γ_0 is developed in different literature [Helfand 82, Pearson 82, Hyun 09, Hyun 11] and could be summarized to the following power law behavior, see also Figure 2.2c:

$$I_{n/1} \propto \gamma_0^{n-1} \; . \tag{2.11}$$

This scaling dependence was used by Hyun et al. [Hyun 09] to introduce a new nonlinear mechanical quantity:

$$^nQ := \frac{I_{n/1}}{\gamma_0^{n-1}} \quad \text{with} \quad \lim_{\gamma_0 \to 0} {}^nQ := {}^nQ_0 \; . \tag{2.12}$$

As a last characteristic the maximum accessible intensity values in the normalized frequency spectrum is deduced from the Ostwald de Waele model with a maximum reachable nonlinearity or strain softening behavior, respectively:

$$\eta(\dot{\gamma}) = \frac{\eta_0}{(\beta \, |\dot{\gamma}|)^c} \tag{2.13}$$

$$\dot{\gamma} = \gamma_0 \omega_1 \cos(\omega_1 t) \; .$$

At maximum shear thinning, c reaches up to 1 and the shear stress can be expressed as follows:

$$\sigma = \eta\dot{\gamma} = \eta_0 \frac{\dot{\gamma}}{(\beta \, |\dot{\gamma}|)^c} \overset{c=1}{=} \frac{\eta_0 \gamma_0 \omega_1 \cos(\omega_1 t)}{\beta \gamma_0 \omega_1 |\cos(\omega_1 t)|} = \frac{\eta_0 \cos(\omega_1 t)}{\beta |\cos(\omega_1 t)|} \tag{2.14}$$

25

which describes a periodic and symmetric step function. The Fourier analysis of the periodic, symmetric step function yields the maximum intensity values [Bronstein 08]:

$$I(\omega_1) \propto \frac{4}{\pi} \left[\sin \omega_1 t + \frac{1}{3} \sin 3\omega_1 t + \frac{1}{5} \sin 5\omega_1 t + ... \right] . \tag{2.15}$$

The maximum intensity for the envelope function is extracted as:

$$\frac{I_{n\omega_1}^{\infty}}{I_{\omega_1}} = I_{n\omega_1}^{\infty} = \frac{1}{n} . \tag{2.16}$$

Please be aware that this is an approximation for the maximum intensity value based on an assumption of $c = 1$ where for example no phase transitions are considered and which is not necessarily reached in experiments.

Phase

The second information of the Fourier analysis is the phase value. Former research was based on the graphical analysis via Lissajous figures [Giacomin 98, Giacomin 93a], which does not exhibit specific values and thus does not quantify nonlinearity. To uniquely quantify the values of the phases the measured time data has to be time-shifted. The reason is that every measurement needs the same "starting point" to gain comparable phases, which is equal to a mechanical trigger of $\sigma(t)$. At any time the shear stress response can be expressed by Equation (2.9). The absolute values of the phases ρ_n are extensive quantities like the absolute intensities I_n. Referring the phases of the higher harmonics to the fundamental phase ρ_1 of the shear stress response an intensive quantity is achieved, comparable with the relative higher harmonics $I_{n/1}$. The time data is shifted by a factor of $-\frac{\rho_1}{\omega_1}$ which allows substituting $t = t' - \frac{\rho_1}{\omega_1}$:

$$\sigma\left(t' - \frac{\rho_1}{\omega_1}\right) = I_1 \cos\left(\omega_1\left(t' - \frac{\rho_1}{\omega_1}\right) + \rho_1\right) + I_3 \cos\left(3\omega_1\left(t' - \frac{\rho_3}{\omega_1}\right) + \rho_1\right) + ... \tag{2.17}$$

$$\sigma\left(t' - \frac{\rho_1}{\omega_1}\right) = I_1 \cos(\omega_1 t') + I_3 \cos(3\omega_1 t' + (\rho_3 - 3\rho_1)) + ...$$

From Equation (2.17) the definition for the comparable relative phases of the higher harmonics can be concluded. It is the harmonic phase of the stress response correlated to the phase of the fundamental peak:

$$\phi_n = \rho_n - n\rho_1 . \tag{2.18}$$

The values of ϕ_n vary between $0\,°$ and $360\,°$ and are independent from the starting point of the shear stress in the time domain, due to the correlation of the higher harmonics phases,

ρ_n in Equation (2.9), to the fundamental phase instead of to a correlation to the shear excitation.

The relative phase of the third harmonic ϕ_3 is determined to possess characteristic values for different types of shear stress responses [Neidhöfer 03, Klein 05]. In the case of strain softening behavior ϕ_3 is measured with $180\,°$, whereas in samples with strain hardening character ϕ_3 possess a value of $360\,°$.

2.2.2 Pipkin diagram

Rheology is divided into the linear and nonlinear regime visualized in the Pipkin diagram as a function of the assigned strain amplitude γ_0 and excitation frequency $\omega_1/2\pi$, see Figure 2.5 [Pipkin 72, Macosko 94, Giacomin 98, Wilhelm 02]. If a shear is applied, the ratio of the characteristic relaxation time τ_{relax} and the flow time t_f of the material define the dimensionless Deborah number:

$$D_e = \frac{\tau_{relax}}{t_f} \, . \tag{2.19}$$

In an oscillatory experiment t_f is the inverse of the applied frequency $\omega_1/2\pi$, resulting in $D_e = \tau_{relax} \cdot \omega_1$. This means the flow is compared to the ability of the material to flow. There are three major areas, which are mainly dominated by D_e:

$$D_e \ll 1, \quad D_e \approx 1, \quad D_e \gg 1 \, . \tag{2.20}$$

The region for $D_e \gg 1$ has a short time of observation t_f compared to the mean relaxation time τ_{relax}, thus stress relaxation is not possible. Below the yield strain γ_{yield} the response is like an elastic solid and obeys Hook's law. Above γ_{yield} the material ruptures or yields into a fluid. For $D_e \ll 1$ the frequency is low and the sinusoidal shearing can be regarded as steady shearing. This means the relaxation times are small compared to the flow times and the material behaves like a Newtonian liquid. When $D_e \approx 1$ viscoelastic behavior is expected. For large γ_0 the nonlinear regime can be reached. With the application of FT-Rheology and its possibility of varying frequency and strain amplitude independently every point in the Pipkin diagram is accessible. Within this thesis the nonlinear viscoelastic regime is the most intensely investigated region.

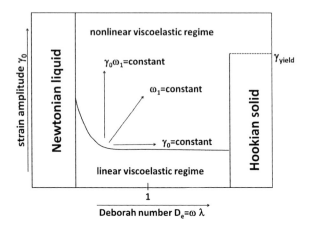

Figure 2.5: Pipkin diagram with its three possible experiments to investigate the nonlinear regime using FT-Rheology [Pipkin 72, Giacomin 98].

2.2.3 Advantage of FT-Rheology

Industrial process conditions, like extrusion or pipe flow, generally exceeds the linear regime and thus the nonlinear oscillatory shear becomes important. Large amplitude oscillatory shear measurements gain system properties in the nonlinear regime and can give, therefore, relevant information for improvements in the process mechanisms. Kallus et al. [Kallus 01] has shown an example where steady shear measurements in the nonlinear regime could not distinguish the properties of different samples. The FT-Rheology was used as complementary characterization and the differences between the samples could be resolved. In research LAOS is used for several systems and different applications. Meins et al. [Meins 11] used FT-Rheology to measure the orientation of block copolymers. The discrimination of linear and entangled polymers via the nonlinear mechanical coefficient Q_0 is done by Hyun et al. [Hyun 09]. As a last example the noninvasive characterization of emulsions with respect to droplet size and interfacial tension is done within this thesis, Section 7.7. FT-Rheology acquires data using a sampling rate much faster than the timescale of rheological experiments but averages them on-the-fly before FT is applied. Thus, the important information is quantified in values of intensity and phase of the higher harmonics. Nonlinear rheology is determined by the onset of measurable higher harmonics contributions. The graphical analysis via Lissajous ($\sigma(\gamma)$) and viscous Lissajous ($\sigma(\dot{\gamma})$)

plots is less sensitive to quantify nonlinearity. A visible influence of nonlinearity on the shear stress response as function of either γ or $\dot{\gamma}$ is observed above 4 % and 2 %, respectively, relative third harmonic contribution. The analysis based on Fourier Transformation yield intensity and phase spectra, which quantify the nonlinear contribution. The sensitivity with respect to detection of nonlinearity was increased by a factor of 100 in comparison to the graphical analysis and in Chapter 3 a further improvement of the signal to noise ratio in the frequency dependent intensity spectrum together with the development of the rheometers will exhibit again a factor of 100 with respect to the lower sensitivity limit of nonlinearity. In the next chapters constitutive equations either based on the mode coupling theory for suspensions, Chapter 6, or on elliptical models for emulsions, Chapter 7, are used to predict the nonlinear behavior of the samples, which are investigated by FT-Rheology.

3. Optimization of the signal to noise ratio

3.1 Influencing parameters

The signal to noise ratio is introduced in Section 2.1 as a parameter, which defines the quality of an FT-spectrum and limits the lowest detectable nonlinear intensity. The Fourier Transform Rheology detects the shear stress in the time domain and applies the Fourier Transformation to achieve a frequency spectrum of intensity and phase or real and imaginary part, respectively. This measurement technique is an oscillatory, rheological investigation method generally used to explore the nonlinear regime of all kind of samples, from polymer melts over almost solid dispersions to liquid fluids [Wilhelm 02]. In the following, the intensity (or magnitude) spectrum is taken as a quantification analysis for the nonlinearity. For comparative reasons the frequency domain is normalized to the intensity of the fundamental peak I_1 appearing at the excitation frequency $\omega_1/2\pi$ to create intense quantities, $I_n/I_1 = I_{n/1}$ at frequencies $n\omega_1/2\pi$ with n being odd multiples of one. The application of FT-Rheology is focused on the detection of higher harmonics, as the peculiarity of this measurement technique is the characterization of the nonlinear regime. Equation (2.11) shows that within the concept of the ^nQ coefficient and the high sensitivity of FT-Rheology the linear regime is defined by vanishing nonlinearities. Thus, the linear regime is achieved for only vanishing deformations and, therefore, never for any real experiment. Nevertheless, it is commonly accepted that linear response can very accurately describe the mechanical response. The linear regime in oscillatory shear tests is assumed to be approximated by the mechanical response with only 0.5 % nonlinearity, $I_{3/1} < 5 \cdot 10^{-3}$, from FT-Rheology [Hyun 09]. The sensitivity, defined by the signal to noise ratio, limits the application of FT-Rheology to detect nonliearities $I_{n/1}$ with $n > 1$. A main part of this thesis deals with the application of FT-Rheology on dilute emulsions to detect $I_{n/1}$ with $n = 3$ and 5, Section 7.7.4. Equation (2.11) predicts intensities of the higher harmonics for each strain amplitude even if γ_0 tends to zero. LAOS experiments of dilute emulsions have shown, that the detection of higher harmonics in the frequency spectrum is limited by S/N. The lower the S/N, the better is the detection of the intensities $I_{n/1}$ at small strain amplitudes. Therefore several adjustments in the measurement of the shear stress in the time domain have been made, to improve the signal to noise ratio.

The history of the large amplitude oscillatory shear (LAOS) experiments over the years shows a continuous development which is for example reviewed in [Hyun 11]. In the beginning the shear stress in the time domain was measured with an external recording device, separated from the actual rheometer control. The analyzing was done with a post processing Fourier Transformation. Nowadays, the new generation of rheometers are equipped with a software for LAOS measurements and Fourier Transformation, respectively. The ARES G2 from TA Instruments is the first new generation rheometer, which merges in the software TRIOS the possibility of on the one hand post processing of the raw shear stress in the time domain and on the other hand on-the-fly Fourier Transformation with direct output of the higher harmonics intensities $I_{n/1}$. For improving S/N different settings can be varied to reduce the influence of the following disturbances.

Mechanical noise

External mechanical disturbances from footsteps and building vibrations are reduced by a rigid and mechanically stable environment beyond the rheometer, which consists for example of a heavy stone table with carpet sheets below the legs.

Electrical noise

Generally electronic disturbances within the investigated frequency range are mostly not random but located at specific frequencies like 50 Hz electrical oscillation from the net plug. Contributions can therefore be avoided by choosing an excitation frequency different from an odd factor of 50 to measure a non-disturbed higher harmonic.

Oversampling

Oversampling is a technique to reduce random noise. It enables the recording of the shear stress with the highest sampling rates, here 50 000 pt/s, and a reduced data volume of a factor of 100 to 1000 is achieved by preaveraging them on-the-fly [van Dusschoten 01]. The TRIOS software is a commercial product and can only be used as black box for the user. But comparative measurements have exposed that oversampling seems to be included by TA Instruments, see Figure 3.2. The amount of points per second, namely the sampling rate, determines the maximum frequency in the spectra defined by the Nyquist-Shannon sampling rate theorem [Kammeyer 02, Schmidt-Rohr 94]. Applying this theorem secures negligible loss of information by sampling the torque with

$$t_{dw} = 1/sr \leqslant 1/2\nu_{max} \tag{3.1}$$

with t_{dw} called dwell time, sr sampling rate and $\nu_{max} = \nu_{Ny}$ Nyquist frequency. The amount of cycles has an influence on the width of the peaks in the frequency spectra. The longer the acquisition time, the more cycles are recorded, which means more points are evenly spaced over the fixed frequency range ν_{Ny} to reduce the frequency interval $\Delta\nu$.

Geometry

Beside the software adjustments, the surface of the geometry has an impact on the minimum measured torque. The relation between shear stress response σ, the radius R of the geometry and the measured torque M is $M = \sigma \cdot \pi R^3/2$ and $M = \sigma \cdot 2\pi R^3/3$ for plate-plate and cone-plate geometries, respectively. Accordingly a larger diameter results in a lower measurable torque limited by increasing inertia forces of the geometry. These inertia forces cannot be balanced by the software after yielding a diameter of 60 mm and additionally results in resonances of the transducer.

3.2 Experiments

In the experiments different rheometers and geometries were compared to achieve the best signal to noise ratio, S/N. The rheometers are strain controlled and of two generations. The old one is entitled ARES G1 for first generation and the new one correspondingly ARES G2. The technical specifications differ for the two rheometers and are described in

Section 11.2.1. With the old device the raw strain and stress data were recorded, Figure 2.2a, via double shielded BNC cables from the back of the instrument. The connection with an ADC card yields a digital periodic signal as output, which was treated with a home written MATLAB® routine subsequently. Accordingly, no direct measurements of the higher harmonics as function of strain amplitude could be achieved, Figure 2.2c. The ARES G2 has the advantage of the incorporated FT-software within TRIOS, which does not only facilitate the measurements but also encounts for friction effects of the geometry to increase the quality of the measured response [Franck 08]. The adjustment for these measurements are the points per cycle and the number of cycles over which the Fourier transformation is applied. In TRIOS, it can be additionally chosen between direct measurements of $I_{n/1}(\gamma_0)$ and the recording of the raw stress data with post analysis via MATLAB®. The S/N can be read for example from the frequency spectra $I_{n/1}(\omega_1)$. The measurement of the strain amplitude dependent higher harmonics, $I_{n/1}(\gamma_0)$, expose the minimum detectable nonlinearity. In the following measurements both illustration possibilities $I_{n/1}(\omega_1)$ and $I_{n/1}(\gamma_0)$ for nonlinear behavior will be shown for different measurement adjustments of a dilute polymer blend of PDMS in PIB (characterization in Section 7.7.1) with a volume fraction of $\Phi_{Vol} = 10\%$. In the following different measurements were applied:

Method A ARES G1 (100FRT for torques from $2 \cdot 10^{-}4\,\text{mNm}$ to $10\,\text{mNm}$) with external record of the shear stress in the time domain for post analysis with a homewritten MATLAB® Routine

Method B ARES G2 (FRT for torques from $50\,\text{nNm}$ to $200\,\text{mNm}$) with external record of the shear stress in the time domain for post analysis with a homewritten MATLAB® Routine

Method C ARES G2 with record of the shear stress in the time domain with the TRIOS software from TA Instruments and post analysis with a homewritten MATLAB® Routine

Method D ARES G2 with record of the strain amplitude dependent nonlinearities $I_{n/1}$. The correlation mode applies the FT on-the-fly over an average of cycles. The software limits the averaging to up to 30 cycles.

3.3 Results for the signal to noise optimization

In Figure 3.1 the comparison of the frequency spectra measured with the ARES G1 and ARES G2 are compared. Both measurements were made with the external recording of the shear stress in the time domain and post analysis with Fourier Transformation. The ARES G2 exhibits for low frequencies as well as for the high frequency spectrum a qualitatively better result with a lower noise level of factor 2.8. Additionally, the fundamental peak at the excitation frequency of 0.1 Hz is more narrow.

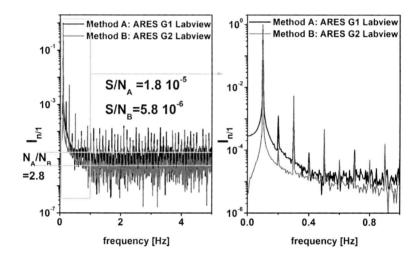

Figure 3.1: Measurement of the intensity spectra via different rheometers of a dilute PDMS/PIB emulsion with a volume fraction of $\Phi_{Vol} = 10\%$, matrix viscosity $\eta_m = 68\,\text{Pas}$ and a viscosity ratio of $\lambda = 4.2$, see Section 7.7.1. Comparison of the higher harmonics intensity in the frequency domain measured at an excitation frequency of 0.1 Hz and a strain amplitude of $\gamma_0 = 2.914$ with the ARES G1 and ARES G2. The ARES G2 shows an overall better result: more narrow fundamental peak and lower noise level of a factor 2.8. The measurement conditions were chosen equally for better comparison: cone plate geometry with angle 0.04° and 50 mm in diameter, 200 points/cycle and averaging over 20 cycles to secure equilibrium state in oscillatory shear.

As a next step the external record of the shear stress via BNC cable is compared with the detection of the shear stress in the time domain with TRIOS at the more sensitive rheometer ARES G2, Figure 3.2. Both raw data sets were post treated with the same home written MATLAB$^{®}$ routine to acquire the frequency spectra. Although the external recorded signal results after the FT in a more narrow fundamental peak, the noise level at high frequencies is reduced by a factor of 40, if measured with TRIOS, see Figure 3.2. The TRIOS measurement exhibit a noise level of $2.5 \cdot 10^{-7}$ compared to the external record with a noise level of $1.1 \cdot 10^{-5}$. The commercially available software can only be used as a black box and, therefore, the improvement in sensitivity cannot explained. It is important that the noise level is already reduced at $0.5\,\mathrm{Hz}$ which improves the detection of $I_{5/1}$ in the small amplitude region, where the intrinsic nonlinearities $^{n}Q_0$ values are determined, see Equation (2.12).

Another influence parameter is shown in the next Figure 3.3. As mentioned in Section 3.1 the frequency interval between two consecutive points reduces with an increasing measurement time or increasing recorded cycles. Additionally an averaging over cycles has an influence on the noise level, described by $S/N \propto \sqrt{n_c}$ for random noise with n being the number of cycles. The averaging of $100\,\mathrm{cycles}$ compared to the averaging of $10\,\mathrm{cycles}$ should result in a reduction of the random noise by a factor of $\sqrt{10} \approx 3.2$. But the measurement exhibits an improvement of a factor of 5.5. In Figure 3.3a the noise level is highlighted with gray lines for a better comparison of the noise levels using different detection methods at an excitation frequency of $0.1\,\mathrm{Hz}$ and a strain amplitude of $\gamma_0 = 3.135$. In Figure 3.3b the correlation mode is compared with the detection of the shear stress in the time domain and post analysis with FT. The record of the time dependent data were averaged over $100\,\mathrm{cycles}$, whereas the correlation mode calculates $I_{n/1}$ of an average of only $10\,\mathrm{cycles}$. The superposition of the scaling law behaviors of $I_{n/1}$ is expressed in $I_{n/1} = a_n \cdot \gamma_0^{-1} + b_n \cdot \gamma_0^{n-1}$ with n being the number of overtone, see Equation (7.31). This nonlinear fit combines the region of dominated by instrument noise with a γ_0^{-1}-dependence at very small strain amplitudes with a γ_0^{n-1}-dependence at increasing strain amplitudes dominated by the nonlinear mechanical response of the sample. The transition between the two scaling dependencies is defined at $\gamma_0 = \sqrt[n]{\frac{a_n}{b_n}}$ and is characteristic for the sensitivity of the instrument. The lower γ_0 the more sensitive is the measurement technique. Additionally the prefactor a_n determines the quality of the measurement. Increasing sensitivity is reflected in a decreasing a_n value. A lower a_n value means a lowered instrument noise expressed in the lowered γ_0^{-1}-dependence. This induces a transition to the γ_0^{n-1}-

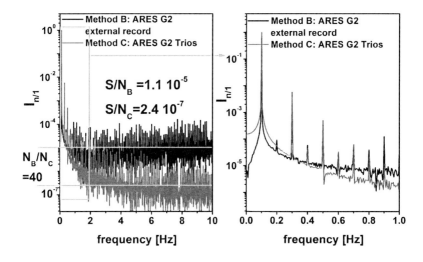

Figure 3.2: Measurement of the intensity spectra from the ARES G2 rheometer as a function of the different recording opportunities described in Method B and Method C for a dilute PDMS/PIB emulsion with a volume fraction of $\Phi_{Vol} = 10\%$, matrix viscosity $\eta_m = 68\,\mathrm{Pas}$ and a viscosity ratio of $\lambda = 4.2$, see Section 7.7.1. The detection of nonlinearity with the TRIOS software results in a reduced noise level with a factor 40 better than via the external recording. The absolute noise level has a value of $2.5 \cdot 10^{-7}$ and is assumed to be the most sensitive spectrum detected so far. The fundamental peak is widened in contrast to the external recorded shear stress. Both signals were post processed with the Fourier Transformation by the home written MATLAB® routine. Both measurements were performed at $0.1\,\mathrm{Hz}$ and $\gamma_0 = 3.135$ in a cone plate geometry with angle $0.04°$ and $50\,\mathrm{mm}$ diameter. The data recording was equal for both measurements with $200\,\mathrm{points/cycle}$ and $20\,\mathrm{cycles}$ for averaging.

dependence, dominated by the nonlinear mechanical response of the sample, at smaller γ_0. The comparison of the correlation mode with the detection of the shear stress in the time domain yield a better sensitivity for the correlation procedure, see Figure 3.3b. The prefactor a_n determined for $I_{3/1}$ is $9.19 \cdot 10^{-6}$ and $2.28 \cdot 10^{-5}$ for method D and method C, respectively. Especially for $I_{5/1}$ the improvement of the correlation procedure is obvious and results in $a_5 = 9.98 \cdot 10^{-6}$ and $7.00 \cdot 10^{-5}$ for method D and method C, respectively. The faster correlation mode is advantageous for the sensitivity but also for samples where

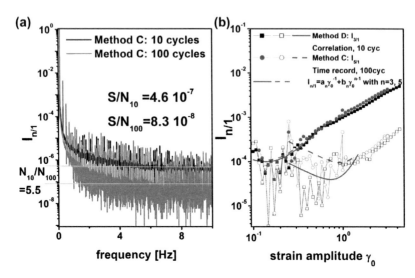

Figure 3.3: Measurement of a dilute PDMS/PIB emulsion with a volume fraction of $\Phi_{\text{Vol}} = 10\,\%$, matrix viscosity $\eta_m = 68\,\text{Pas}$ and a viscosity ratio of $\lambda = 4.2$, see Section 7.7.1. a) With an increasing number of cycles the S/N is reduced by a factor of $\sqrt{n_c}$ with n_c being the number of cycles. A factor of 10 in cycles should reduce the noise level by $\sqrt{10}$. Here, factor of 5.5 is measured. The shear stress is recorded at $\omega_1/2\pi = 0.1\,\text{Hz}$ and $\gamma_0 = 3.135$. b) Strain amplitude dependent measurements of the nonlinearities $I_{3/1}$ (full symbols) and $I_{5/1}$ (open symbols) for the correlation mode in comparison with the measurements of the shear stress in the time domain with post analysis with FT for each strain amplitude, both measured with TRIOS. The quality of the measurement procedure is detected with Equation (7.31) where the prefactor a_n decreases with increasing sensitivity. Here, the correlation mode has lowered a_n values calculated from $I_{3/1}$ and $I_{5/1}$. The excitation frequency is $0.1\,\text{Hz}$ and the geometry is cone plate (0.04°, $50\,\text{mm}$). The points per cycle was for both measurements $200\,\text{points/cycle}$.

evaporation plays a significant role, for example in oil in water emulsions. The correlation mode is limited with respect to the oscillation cycles and here, 10 cycles are chosen in foresight to the future experiments of emulsions with respect to stability of the samples and expenditure of time of the experiment.

As a last improvement the influence on the surface of the geometry will be tested. The larger the surface, the lower is the measurable torque, accordingly the lower is S/N in theory. In experiments three different geometries with diameters of 25, 50 and 60 mm were used to investigate the large amplitude oscillatory shear behavior of PDMS in PIB with a diblock copolymer as compatibilizer. The concentration of the compatibilizer is low enough, that it does not influence the material properties of the raw phases PIB and PDMS, [van Pyuvelde 02]. It is used to vary the interfacial tension as indeed shown in Table 7.1. The 60 mm geometry is a custom product and only available as plate plate type. In a plate plate geometry the shear rate $\dot{\gamma}$ is defined as:

$$\dot{\gamma} = \frac{R \cdot \Omega}{h} \tag{3.2}$$

with R being the radius, Ω the angular velocity and h the gap between upper and lower plate [Macosko 94]. With the radius in the numerator, the shear rate depends on the radius and is therefore not constant throughout the measurement gap. Whereas the cone plate geometry has the advantage of a radius independent shear rate:

$$\dot{\gamma} = \frac{\Omega}{\beta} \tag{3.3}$$

where β is the angle of the cone. The onset of the nonlinearity $I_{n/1}(\gamma_0)$ for $n = 3, 5$ is detected by the superposition of the two scaling regions in Equation (7.31). For the cone plate geometries with increasing diameter the γ_0^{-1}-dependence for $I_{3/1}$ is detected to be at $\gamma_0 = 0.16$ and 0.11 for 25 mm and 50 mm, respectively, see Figure 3.4 and Table 3.1. The parallel plate with the largest diameter of 60 mm possess the transition at the highest strain amplitude at $\gamma_0 = 0.3$. For $I_{5/1}$ the following strain amplitudes are measured: $\gamma_0 = 0.73, 0.70$ and 0.75 for 25 mm, 50 mm and 60 mm, respectively. The 60 mm plate plate geometry is less sensitive than the cone plate geometries, where an increasing diameter increases the sensitivity. Table 3.1 summarizes the determined prefactors a_n and b_n with the corresponding γ_0 values. The distinct shoulder in $I_{5/1}$ in Figure 3.4 increases in height with increasing diameter of the geometry. The differences could be due to the different loads as well as due to the different sensitivities of the geometries. A deviation of the height of the shoulder between plate plate and cone plate geometries could be explained with the difference in the shear rate definitions in Equation (3.2) and (3.3).

In conclusion it could be summarized that the ARES G2 has an improved sensitivity of a factor of 2.8 with respect to the ARES G1 under optimized conditions with the same detection method via the external recording (method A). It was shown that the correlation mode (method C) yields a factor of 40 in sensitivity in comparison to the record of the

Table 3.1: Comparison of the sensitivity of three geometries with different diameters measured with a dilute PDMS/PIB emulsion with compatibilizer PIB-b-PDMS. The geomtries are a cone plate geometry (CP) with 25 and 50 mm and a plate plate geometry (PP) with 60 mm. The corresponding strain amplitude dependent $I_{n/1}(\gamma_0)$ measurements are shown in Figure 3.4. The nonlinear fit $I_{n/1} = a_n \cdot \gamma_0^{-1} + b_n \cdot \gamma_0^{n-1}$ (Equation (7.31)) is used to determine a_n and b_n and the corresponding strain amplitude at $\gamma_0 = \sqrt[n]{\frac{a_n}{b_n}}$ where the scaling region, dominated by the instrument noise, goes over into the scaling region, dominated by the nonlinear response of the material. Low values of a_n or γ_0 reflect a high sensitivity with respect to the measurement of nonlinearity $I_{n/1}$.

geometry	a_3	a_5	$b_3 =^3 Q_0$	$b_5 =^5 Q_0$	$\gamma_0 = \sqrt[n]{\frac{a_3}{b_3}}$	$\gamma_0 = \sqrt[n]{\frac{a_5}{b_5}}$
CP 25 mm	$1.55 \cdot 10^{-5}$	$2.40 \cdot 10^{-5}$	$3.73 \cdot 10^{-3}$	$1.12 \cdot 10^{-4}$	0.16	0.73
CP 50 mm	$0.43 \cdot 10^{-5}$	$2.26 \cdot 10^{-5}$	$3.38 \cdot 10^{-3}$	$1.31 \cdot 10^{-4}$	0.11	0.70
PP 60 mm	$4.28 \cdot 10^{-5}$	$4.11 \cdot 10^{-5}$	$1.51 \cdot 10^{-3}$	$1.68 \cdot 10^{-4}$	0.31	0.75

raw shear stress in the time domain (method B). With method C an optimized spectrum with S/N of $\approx 10^7$ was measured and is assumed to be the maximum in sensitivity with respect to mechanical characterization via FT-Rheology with the state of the art rheometers. Giacomin et al. [Giacomin 98] shows the first spectrum of a nonlinear mechanical measurement with $S/N = 10^2$. Thus, the hardware and software developments of the rheometers, see Appendix 11.2, yield a factor of 10^5 in sensitivity. The construction of a custom plate plate geometry (see Appendix 11.3) with 60 mm in diameter could not achieve the expected improvement in sensitivity due to increasing surface compared to the standard cone plate geometry with 50 mm diameter. The increasing diameter in the cone plate geometry from 25 to 50 mm lowers the onset of nonlinear scaling behavior with γ_0^2-dependence for $I_{3/1}$ from $\gamma_0 = 0.16$ to 0.11. Thus, for the following measurements the correlation mode with a cone plate geometry of 50 mm is used for emulsion rheology to guarantee maximum sensitivity and constant shear rate throughout the whole sample.

3.4 Maximum detectable higher harmonics

Many samples show under LAOS a high amount of higher harmonics. Therefore the measurement parameters were optimized to create highest sensitivity in combination with a

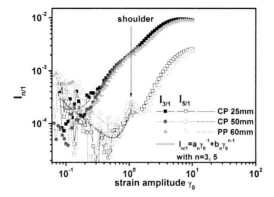

Figure 3.4: Measurement of the strain amplitude dependent $I_{n/1}(\gamma_0)$ with method D of the PDMS/PIB emulsion with 0.7 % PIB-b-PDMS diblock copolymer as compatiblizer, see Section 7.7.1. The material properties are: volume fraction of $\Phi_{\mathrm{Vol}} = 10\,\%$, matrix viscosity $\eta_m = 68\,\mathrm{Pas}$ and viscosity ratio of $\lambda = 4.2$. Increasing diameters in geometries increase the sensitivity with respect to the minimum measurable torque and the same type of geometry. The detection of the strain amplitude to the corrsponding transition from dominated noise to dominated sample dependency is determined by Equation (7.31). The values can be found in Table 3.1.

maximum number of nonlinearities $I_{n/1}$ with n being maximum. In an experiment a commercial sample with high volume fraction of the dispersed phase, w/o-2 (see Section 7.7.1), is sheared in a cone plate geometry. The interactions between the droplets cause a network structure, which is distorted under LAOS and generates plenty of higher harmonics. The 50 mm cone plate geometry was used for measurements in the ARES G1 and ARES G2. The two rheometers of different generations with different sensitivity limits are compared in Figure 3.5a and b. The ARES G2 has an improved torque resolution, which is confirmed in the increased maximum higher harmonic from 147th to the 189th. Both measurements were measured at an excitation frequency of 0.1 Hz and a strain amplitude of $\gamma_0 = 870$ in absolute values with a sampling rate of 50 points/s over 10 cycles. The shear stress in the time domain was externally record and postpone Fourier transformed. In Figure 3.5c a new world record of higher harmonics with the ARES G2 was measured. The strain amplitude could be increased to $\gamma_0 = 3000$ and 289 overtones could be measured, Figure 3.5c.

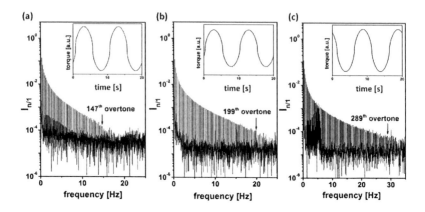

Figure 3.5: LAOS measurements with a commercial w/o-emulsion with method A. a) Measurement of the frequency spectrum with the ARES G1 yields a maximum amount of 147 higher harmonics. b) Comparative measurement as in a) with the more sensitive ARES G2 as specified by the manufacturer. The excitation frequency is 0.1 Hz and the strain amplitude $\gamma_0 = 870$. The cone plate geometry spans 50 mm in diameter. The sampling rate is 50 points/s and the shear stress was recorded over 10 cycles. The number of maximum higher harmonics is increased from 147 to 189 overtones. In c) the new record of higher harmonics, is measured with the ARES G2. Here the strain amplitude is maximized to $\gamma_0 = 3000$ in absolute values resulting in 289 overtones.

3.5 Comparison between the ARES G2 and the new Discovery Hybrid Rheometer DHR2

Rheology measures the mechanical properties of the here investigated soft materials. The sensitivity is based on the setup of the rheometers. The improvements of gear and bearing of the motor and of the transducer increase the quality of the stress signal. Within the last period of this thesis, a new rheometer was developed and commercialized as DHR2 by TA Instruments. The DHR2 is a stress controlled rheometer with the opportunity of strain controlled measurements. The torque resolution as specified by the manufacturer is as low as 2 nNm, which is a factor 25 better compared to the resolution of the ARES G2 (50 nNm). The influence on the signal to noise ratio was already described above for the ARES G1 and ARES G2 and will be shown for measurements with the new Hybrid Rheometer DHR2 in the following. Figure 3.6a shows the measurement of w/o-2 with the ARES G2 in comparison to th DHR2 in Figure 3.6b with the same measurement conditions. These have been actually a parallel plate geometry of 40 mm diameter, an excitation frequency of 0.1 Hz, a strain deformation of 200 in absolute values and a recording of the shear stress in the time domain over 10 cycles. The sampling rate for the ARES G2 was set to 50 points/s. The DHR2 has a predefined sampling rate of 1000 points/s, which was smoothed by 51 points in post processing. The smoothing is achieved by a convolution of the shear stress in the time domain with a rectangular function containing a length of 51 points. The noise level in Figure 3.6 is reduced about a factor of 3.8, which confirms the increased sensitivity of the DHR2. But the maximum overtone is lowered in Figure 3.6b, due to irregular intensities at high frequencies. As seen in Figure 3.6c the intensities of the higher harmonics are comparable between both instruments. The intensities up to $I_{9/1}$ have a standard deviation of below 0.2 %.

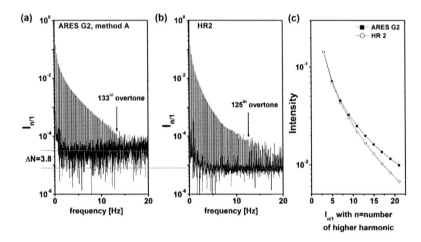

Figure 3.6: Measurement of the magnitude spectrum with (a) the ARES G2 and (b) the DHR2. The sensitivity in torque resolution of the DHR2 is a factor 25 lower than the torque resolution in the ARES G2 as specified by the manufacturer. The excitation frequency $\omega_1/2\pi$ is $0.1\,\mathrm{Hz}$ and the strain amplitude $\gamma_0 = 200$ in absolute values. The noise level is reduced about a factor of 3.7 in the measurement with the more sensitive DHR2. c) Comparison of the intensities of the first 10 odd higher harmonics $I_{n/1}$ with $n = 3$ to 21. The intensity of $I_{3/1}$ is the same for both measurements and the intensities up to $I_{9/1}$ have a standard deviation of below $0.2\,\%$. The 21st overtone deviates about $0.22\,\%$ from the ARES G2 measurement to the measurement with the DHR2.

4. Stabilization of dispersions

Dispersions are thermodynamically instable systems consisting of two immiscible phases. The state of aggregation of the two phases divides the dispersions into different groups. Within this thesis three types of dispersions are investigated: polymer colloids (Chapter 6), emulsions (Chapter 7) and foams (Chapter 9). The different destabilization procedures, as they take place in a dispersion which is not in an equilibrium state, are shown in Figure 4.1 [Dörfler 02, Butt 10, Schuchmann 05]. The evolution of a dispersion without stabilization

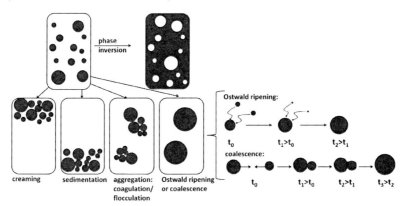

Figure 4.1: Schematic picture of the different destabilization effects in a dispersion [Dörfler 02, Butt 10, Schuchmann 05].

and additional distinct density difference between dispersed phase and dispersion medium, shows creaming or sedimentation. Flocculation, Ostwald ripening and coalescence of the dispersed phase is mainly dependent on the kinetic stabilization via steric or electrostatic stabilization as figured out in Sections 4.2 and 4.1, respectively.

Experiments of dispersions are dependent on the stability of these systems. The stability is determined by the attractive forces which cause aggregation either by flocculation or by coagulation [Piirma 92]. Flocculation and coagulation are first defined by Napper [Napper 83] and describe reversible and irreversible aggregation, respectively. Due to Brownian motion the particles stay in motion and collide occasionally. These collisions are enforced by attractive interactions and can be prevented by repulsion forces, which increase stability. Attractive forces are primarily due to particle-particle interactions and can be counterbalanced by electrostatic or polymeric stabilization. First one is based on an electric double layer described in the following Section 4.1 and the latter one on a steric or a depletion stabilization, Section 4.2. Attractive forces have different origins and the dimension of the potential energy decreases with the inverse sixth power of the radius, $E_A \propto -\frac{1}{R^6}$ like van der Waals interactions. Attractive forces can be due to permanent dipole-permanent dipole (Keesom) forces, permanent dipole-induced dipole interactions (Debye) and London dispersion forces based on transitory dipole-transitory dipole forces [Piirma 92, Atkins 07]. Repulsive energies are in principle low range interactions, because they decrease with R^{-12}.

Generally surfactant molecules consists of a hydrophilic head group and a lipophilic part like hydrocarbon chains, see Figure 4.2. They can stabilize a dispersion with two mechanisms: electrostatic stabilization or steric stabilization. Electrostatic stabilization is mainly applied to oil in water emulsions (o/w-emulsion), where the charged head groups look into the hydrophilic phase, Figure 4.2b. In colloidal systems with charged molecules at the interface of the dispersed phase and dispersion medium, the electrostatic repulsion energy also depends on the thickness of the electric double layer. In water in oil emulsions (w/o-emulsion), steric stabilization takes place using the inverse orientation of the surfactant molecules, which thus spread the hydrophobic hydrocarbon chains into the oil phase, Figure 4.2a and Figure 4.5.

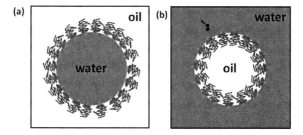

Figure 4.2: Schematic picture of a w/o-emulsion (a) and o/w-emulsion (b) with surfactant molecules at the interface of the immiscible phases. The hydrophilic part is oriented with the hydrophilic head group into the water phase and the lipophilic part, for example hydrocarbon chains, into the oil phase.

4.1 DLVO theory

Within this thesis the main points for the stabilization of emulsions will be stated. They are directly related to the DLVO-theory. From 1939 until 1945, the scientists Derjaguin and Landau from Soviet Union and at the same time Overbeek and Verwey from Netherland created the bases of the DLVO theory for dispersed colloids [Dörfler 02]. As for dispersions, the stabilization is based on attractive (E_A) and repulsion energies (E_R) which are summed up to the total potential energy E_t:

$$E_t = E_A + E_R .\tag{4.1}$$

In stable systems it is assumed that the repulsive forces between the charges at the electric double layer of adjacent particles is balanced by the attractive van der Waals interactions [Atkins 07]. The electrostatic repulsion energy, E_R, is expressed by:

$$E_R = ze\psi_0 \exp\left(-\chi_D d_A\right)\tag{4.2}$$

where z is the number of elementary charge, e the elementary charge, d_A the distance between two particles, ψ_0 the potential at $d_A = 0$ and χ_D has the dimension of an inverse length with $\frac{1}{\chi_D}$ being the Debye length. The Debye length has the dimension of inverse length scale and gives an estimation about the decay of the electrostatic potential [Larson 99, Cosgrove 05]. The ions at the interface of the particles exert attractive and repulsive interactions towards the surrounding ions and counter ions. The presence of these charges screen the potential of the central ions. The Debye length is a characteristic length scale, that describes the screening effect of the charges onto the distance of the electrostatic potential of the central ions. This means the higher the concentration of ions in the dispersion is, the stronger is the screening effect and, therefore, the shorter is the Debye length. Charges on a surface of a droplet appear of different reasons, like

- dissociation of ionic groups on the interface

- adsorption of ions from the continuous phase on the droplet surface.

These charges build two different potential areas with respect to the distance from the interface. Next to the droplet surface the ions are fixed generating an electric potential. This fixed layer is followed by a diffusive (mobile) layer Figure 4.3 [Dörfler 02].

For a w/o-emulsion the thickness is estimated to be 1 to $5\,\mu m$ because the continuous oil phase contributes no free charges to screen the electrostatic potential coming from the particle interface. In contrast to this a o/w-emulsion possess a thickness of only $10^{-3}\,\mu m$ due to the screening effect of free charges in the water phase [Dörfler 02]. If

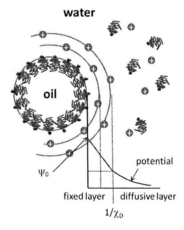

Figure 4.3: Double layer structure of an oil droplet with an anionic activated surfactant on the interface oil/water. Next to the droplet surface the ions are fixed generating an electric potential. This fixed layer is followed by a diffusive (mobile) layer [Dörfler 02].

the thickness is reduced beyond a critical minimum the particles can approach each other with less repulsion until they coalesce. This means a flattening of the repulsive energy curve resulting in a total energy with decreasing $E_{T,\max}$ until the right arm of E_T shows a totally negative potential, see Figure 4.4b. The repulsion can be reduced by adding an electrolyte, which shields the free charges. The diffusive double layer shrinks and the emulsion becomes unstable. The attractive potential between two atoms is also called

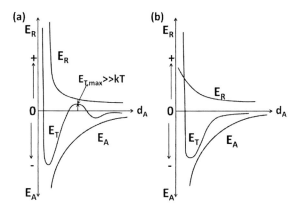

Figure 4.4: Energy versus distance d_A as as sum of repulsion energy, E_R, and attractive energy, E_A. (a) A stable dispersion has a total energy $E_T = E_A + E_R$ with a maximum $E_{T,\max}$ in the positive region of the potential, where $E_{T,\max}$ is higher than the thermal energy $E_{T,\max} \gg kT$. (b) An instable dispersion possess a flattened repulsion curve due to a decreased electric double layer, which reduces the maximum to zero. Thus, the energy barrier for coagulation and flocculation is vanished.

London-attractive-potential defined by $E_A = -3\alpha^2 h\nu_0/4r^6$, where α is the polarizability, ν_0 the critical frequency of the atom and r the radius of the atom [Dörfler 02]. But for the attractive interaction between two spheres the range of interaction changes to:

$$E_A = -\frac{A_H r}{12(R - 2r)} \quad \text{for small distances} \tag{4.3}$$

$$E_A = -\frac{2r^2 A_H}{3R^2} \quad \text{for long distances, } R \gg 2r \tag{4.4}$$

where A_H is the Hamaker Konstant, r the radius of the spheres, R the distance between the centres of the neighbored spheres [Dörfler 02]. This brief introduction into electrostatic stabilization will not be discussed in more detail, because within this thesis mainly polymeric stabilization is used. Only the commercial w/o-emulsions in Section 7.7.1 are stabilized by electrostatic repulsion.

4.2 Polymeric stabilization

Polymeric stabilization is a steric stabilization effect which prevents the droplets to coalesce due to adsorbed or attached macromolecules at the interface [Piirma 92]. It is reasonable that amphipathic block copolymers, with one block soluble in the dispersion medium and the other soluble in the dispersed phase, show an advanced tendency to attach at the interface, which provides steric stabilization with the polymeric chains. Diblock copolymers, therefore, are also called compatibilizer. The macromolecule must adsorb strong enough to the interface so that a collision with an adjacent particle does not cause desorption. A complete coverage of the surface, for example, prevents the escape of the compatibilizer. Figure 4.5 points out the effectiveness of a steric stabilization. Flocculation, see Figure

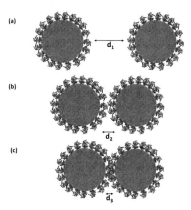

Figure 4.5: Steric stabilization by the chains rising into the dispersion medium. At a distance d_1 no interaction is observed, whereas for a decreasing distance the chains penetrate each other at d_2 until compression appears at d_3.

4.1, is another destabilization effect based on coagulation of dispersed particles. A stable system has a positive Gibbs free energy of flocculation ΔG_f, where ΔG_f is defined by an enthalpy and entropy term:

$$\Delta G_f = \Delta H_f - T\Delta S_f \qquad (4.5)$$

where both can be the primary determination parameter for different stabilization effects:

- enthalphic stabilization occurs, if ΔH_f is the driving term to yield a positive free energy of flocculation

- entropic stabilization, if ΔS_f secures a positive ΔG_f by a larger negative value than the enthalpic term

- enthalpic-entropic stabilization occurs, if both terms help to stabilize, which means a positive ΔH_f and a negative ΔS_f, respectively.

If two adjacent particles, covered with chains rising into the dispersion medium, approach each other until the chains penetrate and interact with each other, the free energy ($\Delta G = \Delta H - T\Delta S$) is increased due to loss of configurational entropy $-\Delta S$. The increase in free energy effects repulsion and thus stability.

Dissolving polymers demands a good solvent where attractive forces are negligible. In a good solvent the polymer chains or polymer network shows swelling like it is shown in Figure 6.5a for the poly(N-isopropylacrylamide) shell. In contrast to this bad solvents cannot compensate the attractive forces and the dissolved polymer chains or polymer network collapses as it is shown in Figure 6.5b. Theta solvents balance attractive and repulsive forces between polymeric chains and cause minimized interaction between the polymer and the solvent molecules, [Lechner 03]. If the solubility of polymeric chains is reduced below the theta solvent, dispersions with a polymeric compatibilizer are unstable. Thus, in a non-solvent, flocculation occurs, which can be reversibly redispersed by adding a good solvent. This effect is used to gain reversibility of flocculation. The theta solvent is equivalent with a theta temperature, because interactions are temperature dependent. The temperature dependent stability of polymer colloids is seen for example in Chapter 6. Varying the temperature effects both the enthalpic and entropic contribution of ΔG_f. If the temperature changes, the extension of the sterical layer in Figure 4.5 is varied. This in turn influences the excluded volume, which effects the solubility (ΔH_f) and changes the possibility of penetration (ΔS_f). Within the colloidal dispersions, Chapter 6, a polymer network of poly(N-isopropylacrylamide), is chemically grown as a shell onto the surface of the polystyrene core, see Figure 6.4. It is said, that an electrolyte balances the charges on the surface of the shell, thus electrostatic stabilization can be excluded. An increase in temperature causes a shrinkage of the thickness of the polymer shell until a critical temperature of 32 °C [Siebenbürger 09]. Above the critical temperature, the polymer collapses and destabilization takes place, because the excluded volume decreases, which means a decreased solubility, which is expressed in a decreasing positive enthalpic term of flocculation.

Dispersions using a non-ionic steric stabilization are insensitive to the occurence of electrolytes. This is no longer valid, if a combination of steric and electrostatic stabilization is chosen, for example, using polyelectrolytes as compatibilizer between the dispersed phase and dispersion medium. A polyelectrolyte consists of polymeric chains loaded with charges and combines therefore both stabilization techniques [Piirma 92]. Within this thesis steric stabilization is chosen for the preparation of the dilute model blends in Section 7.7.1, where a block copolymer is settled at the interface to mediate between the immiscible phases with the corresponding blocks.

5. Surface and Interfacial tension

Surface tension or force is defined as the force acting on the contact area between a liquid or solid phase with a gaseous phase, mostly air, see Figure 5.1. Interface phenomena take place between two liquid phases, like in an emulsion or between a liquid and a solid, like in a suspension. In the literature the nomenclature for surface and interfacial tension are frequently mixed up [Howe 97, Israelachvili 92, Cosgrove 05]. Within this thesis the used nomenclature depends on the aggregate state of the phases like they are defined above [Butt 10]. In the following the processes are described for interfacial phenomena, but in general the same are also valid for surface processes [Butt 10].

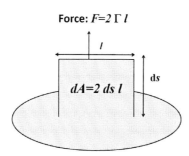

Figure 5.1: Schematic picture of interfacial tension and energy based on Equation (5.1) [Atkins 07]. The force has a factor of two due to the double surface of a liquid lamella [Dörfler 02].

For further derivations, the definition of interface energy and tension are given. The interfacial energy is the minimum amount of work, W_{min}, required to create a new unit area of the interface, A. Interfacial tension is the force, F, acting at a right angle to a line of the unit length, l, in the surface of a liquid [Atkins 07]. Figure 5.1 visualizes the definitions in Equation (5.1):

$$W_{\min} = F \cdot ds = \Gamma \cdot l ds = \Gamma dA \,. \tag{5.1}$$

A thermodynamic derivation of interfacial tension is based on the change of the Gibbs energy, dG:

$$: dG = -SdT + \sum \mu_i dN_i + \Gamma dA \tag{5.2}$$

assuming a constant pressure in both phases, which is at least given for planar interfaces. The pressure difference in spherical droplets will be discussed below. The last term on the right hand side of Equation (5.2) includes the interfacial tension Γ. The interfacial tension is defined as the increase in the Gibbs energy with increasing area at constant T, p and N_i (temperature, pressure and number of particles of the substance i, respectively) [Howe 97, Cosgrove 05]:

$$\left(\frac{\partial G}{\partial A} \right)_{T,p,N_i} = \Gamma \,. \tag{5.3}$$

With this thermodynamic definition, the well known phenomenon of interfacial tension, which is also relevant for this work, can be easily explained. If a system at constant p, T and N_i is considered, the Gibbs energy decreases, $-dG$, if the area decreases, $-dA$. From geometrical point of view, spheres have the smallest surface to volume ratio, which minimizes the Gibbs energy [Butt 10]. Therefore dilute emulsions, like investigated in Section 7.7.1, have spherical droplets at rest, which can be deformed to ellipsoids, if large amplitude oscillatory shear is applied. Equation (5.3) additionally explains, why stabilization of the spheres is necessary to avoid coalescence. Instead of several particles where each droplet creates a new interface, the Gibbs energy can be reduced, if a phase separation takes place, e.g. coalescence, to create two distinct phases with a single interface area. If the dispersed liquid builds spheres, a pressure difference between inner and outer phase is generated. Laplace explored this important fact and quantified it with Equation (5.4):

$$p_{\mathrm{in}} = p_{\mathrm{out}} + \frac{2\Gamma}{R} \tag{5.4}$$

where R is the radius of the sphere. Now it is self-evident that a flat surface, arising by phase separation as a consequence of coalescence, is also enforced by compensating the pressure difference. Equation (5.4) additionally explains the driving force for Ostwald ripening, see Figure 4.1 [Cosgrove 05], which reduces the curvature by growth of the bigger droplets at the cost of the volume of smaller droplets.

5.1 Effect of a surfactant

To explore the effect of a surfactant on the stability of a dispersion, the molecular point of view is regarded [Butt 10]. In a liquid as well as in a solid the molecules attract each other to build the condensed phase. In the bulk phase each molecule is surrounded by adjacent equivalent molecules, whereas at the surface the number of adjacent molecules of the same species is reduced. This imbalance increases the energy and creates an instable condition. The surface energy or interfacial energy, respectively, is the work needed to transport a molecule from the bulk phase to the surface. The function of a surfactant is the decreasing of the energy, i.e. of the interfacial tension, Equation (5.1). The surfactant molecules act as a phase intermediator, which means they are settled at the interface with an affiliation to both phases. The reduction of the interfacial tension, Γ, also reduces the Laplace pressure of Equation (5.4). In practice different kind of surfactants are used. Generally amphiphilic surfactants with a hydrophilic and hydrophobic part are useful for both w/o- and o/w-emulsions. This stabilization can be found in the commercial emulsions in Section 7.7.1. When two immiscible polymers are blended to create an emulsion, Section 7.7.1, a diblock copolymer is added as compatibilizer to generate a lower interfacial tension. A diblock copolymer consists of two chemically linked polymers with different chemical structure. In dispersion stability they are often used as compatibilizer like already mentioned in Section 4.2. Diblock copolymers preferably settle at the interface where the polymer chains reach into the chemically related bulk phase to reduce interfacial tension.

5.2 Measurement techniques

In the literature different types of measurements for surface or interfacial tension can be found [Dörfler 02, Butt 10, Cosgrove 05]. A few of them shall be mentioned, where the relevant will be presented in more detail. The wire bracket method from Lenard is a dynamic method and measures the force at a lamella of a liquid, where the area of the lamella is constantly increased against the attempt of the liquid to minimize the surface area, see Figure 5.2a [Dörfler 02].

The capillary rising method, Figure 5.2b, is a static measurement method and uses the adhesion forces of a liquid in which a thin capillary is immersed [Israelachvili 92, Butt 10]. The fluid will rise the walls of the capillary until a specific height is reached, which in turn is used for the calculation of the surface tension [Dörfler 02].

Another method is the measure of the pressure in a capillary by the size of the appearing bubble Figure 5.2c. This method is a dynamic method and consists of a capillary immersed in a liquid. Due to the rising of the liquid along the walls, the gas at the end of the capillary is pressed into a bubble, which accordingly increases its size from d_1 to d_3 with increasing height h of the liquid. The radius of the liquid contains the information about the surface tension, Equation (5.4), if the hydrostatic pressure is considered to include the information about the depth of immersion [Dörfler 02].

The ring tensiometer of Du Noüy is comparable to the first mentioned method of Lenard, see Figure 5.2d. It measures the force necessary to remove a ring with thin defined diameter out of a liquid.

The static perpendicular plate method of Wilhelmy measures the force induced by a thin plate submerging into a liquid, see Figure 5.3b. The liquid spontaneously wets the surface and climbs upwards the plate walls, which causes a force pulling the plate downwards [Butt 10]. To measure an interfacial tension, the second liquid phase has to be added on top of the first phase until the Wilhelmy plate is completely covered, see Figure 5.3b. For calculation of the surface and interfacial tension, respectively, a contact angle θ close to zero is considered negligible and facilitates the measurement for wetting liquids:

$$F = U\Gamma cos\theta \quad \text{with } \theta < 5° \quad \Gamma = \frac{F}{U} \tag{5.5}$$

where U is the defined perimeter of the Wilhelmy plate and F the measured force. Together with the Wilhelmy plate a new setup for measurements of the interfacial tension with a rheometer is built, see below in the experimental Section 5.3.

The pendant drop measurement analyzes the droplet shape of a downwards hanging droplet from the end of a needle, see Figure 5.2e. An important assumption for this static method is, that the droplet is not in motion and viscosity and inertia do not play a role, whereas surface tension and gravity form the shape of the droplet [Butt 10]. Advances in computational methods pushed the droplet shape analysis. In principle two forces are opposed and

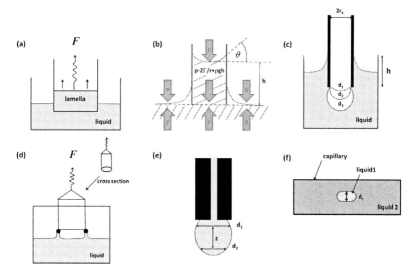

Figure 5.2: Schematic picture of the different surface and interfacial tension measurement methods: a) wire bracket method from Lenard, b) capillary rising method, c) meaure of the pressure in a capillary, d) cross section of Du Noüy ring and e) pendant drop method, f) spinning drop method [Dörfler 02, Butt 10, Cosgrove 05].

influence the droplet profile. The pressure across the curved surface is given by the Laplace Equation (5.4) and is counterbalanced by the gravitational force pulling the droplet down:

$$\Gamma \left(\frac{1}{d_1/2} + \frac{1}{d_2/2} \right) = \Delta p_0 + \rho g z \tag{5.6}$$

where $d_1/2$ and $d_2/2$ are two different radii of the curved surface, which cause a pressure drop Δp across the surface. The pressure difference Δp_0 is measured between a reference plane and an additional pressure arising from the gravitational component at any point on the surface due to the density difference ρ between the droplet and the continuous phase [Cosgrove 05]. The distance z from the reference plane is manually defined in the software, generally. Computational software uses a complex set of first order differential equations to solve numerically the complex analysis of the droplet profile yielding information about the surface or interfacial tension. In Section 7.7 the interfacial tension values of the polymer blend without compatibilizer were measured via pendant drop. The Wilhelmy method failed due to the high viscosities of the raw material.

As a last method the spinning drop measurement shall be explained as another example of a quasidynamic method to measure interfacial tensions, see Figure 5.2f. This method is applicable to systems with a very low interfacial tension, and was, therefore, used to determine the interfacial tension of the commercial emulsion investigated in this work. Within this work, the w/o-emulsion was centrifuged with $20\,000$ rpm ($394\,000$ m/s^2) for 300 min to achieve phase separation. Afterwards the oil phase was separated from the remaining gel like phase. For the interfacial tension the oil phase was measured against distilled water. A very low interfacial tension of 0.09 mN/m was measured. Other interfacial tension measurements failed because of the less precise resolution. The explicit setup of the spinning drop tensiometer is explained in [Dörfler 02]. The main principle of the setup is a rotating droplet in a tube filled with a second liquid of higher density. The high rotation of the tube deforms the sphere to a cylinder due to centrifugal forces, which are opposed by interfacial forces. Further parameters for the analysis are the diameter of the cylinder, d_c, the density differences of the two phases a and b, ρ_a and ρ_b, as well as the angular frequency ω, reaching values of up to 200 Hz:

$$\Gamma = \frac{1}{4}d_c^3\omega^2(\rho_a - \rho_b)\,. \tag{5.7}$$

The spinning drop method can measure interfacial tensions from 10^{-6} mN/m up to 100 mN/m [Dörfler 02].

5.3 Experiments

As part of the characterization of emulsions, the interfacial tension is an important parameter. The Wilhelmy plate method is a static measurement with a simple feasability setup, as described above. Within this thesis, the ARES G2 was used to determine the interfacial tension in combination with a Wilhelmy plate. The ARES G2 allows the measurement of the axial force, while moving the upper geometry upwards or downwards. In an adjusted setup, instead of an upper geometry, the Wilhelmy plate was attached to the rheometer. The lower geometry is replaced by a cup with a diameter two times larger than the width of the Wilhelmy plate to secure a homogeneous surface around the plate, Figure 5.3. The Wilhelmy plate itself is a commercially available plate of platin iridium alloy with dimensions of 19.9 x 0.2 x 10 mm in l x w x h as specified by the manufacturer. The platin iridium alloy can be cleaned by annealing the plate until a homogeneous red heat is achieved. Thus, a clean surface is assumed for the surface and interface measurements. The perimeter of $U = 40.2$ mm is entered into Equation (5.5). The instrument specifications of the ARES G2 define the minimum measurable axial force to 0.001 N, where in the measurements the change of axial force ΔF is measured. The theoretical minimum interfacial tension is therefore calculated to $\Gamma = \frac{\Delta F}{U} = 25$ mN/m, Equation (5.5).

To use the commercially available Wilhelmy plate with the ARES G2 a special sample holder, see Appendix 11.3, was constructed, Figure 5.3c. The sample holder is mounted like the standard geometries for rotational experiments and tightens the Wilhelmy plate with a screw. A roundhole shaft of about 2 cm leads the Wilhelmy plate in perpendicular direction onto the surface of the liquid. If the rising or setting of the plate is too fast, the jump of the axial force is falsified, as well as if the velocity is too slow. The temperature of the cup is controlled via a Peltier system. The sampling rate of 5 mm/s of the axial force as function of time, is high enough to neglect loss of information. Post averaging of 51 points reduces the noise and facilitates the analysis.

5.3.1 Surface tension

As a primary measurement the surface tension of water and a commercial silicon oil (PDMS) with $M_n = 5$ kg/mol (AK100 from Wacker-Chemie GmbH) were measured and compared with the literature value, [Atkins 07], and safety data sheet from Wacker, respectively. The original measurement with the Wilhelmy plate, does not move the Wilhelmy plate, while it is immersed in the liquid. Here, the Wilhemy plate is either moved downwards, from air into the liquid or vice versa from contact with the liquid (only about 1 mm immersion) upwards until the contact with the liquid phase is lost. Since this measurement technique is newly invented, both possibilities will be shown in the experiments. In both cases the axial force suddenly changes if the bottom of the plate touches the surface or looses contact with the liquid, respectively, Figure 5.4. The surface tension is calculated via Equation (5.5), where F is replaced by ΔF, the change in axial force induced by the change of the molecules pulling the plate downwards. The values from the upward measurements agree better with the reference values, than the surface tension values determined by downward movement of the Wilhelmy plate, see Table 5.1. The overshoot

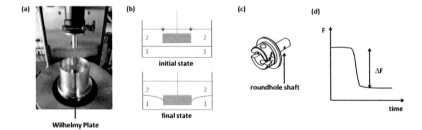

Figure 5.3: a) Photo of the adjusted setup of the ARES G2 for interfacial tension measurements using a Wilhelmy plate and the axial force measurement. b) Schematic setup for interfacial tension measurements. The plate is placed right above the interface and should be covered completely with the second phase of lower density. In the experiment the Wilhelmy plate is lowered into phase one with the higher density. c) Schematic picture of the sampleholder where the head is mounted in the rheometer like a standard geometry for rheological experiments. A roundhole shaft, where a screw thightens the Wilhelmy plate, leads the Wilhelmy plate in a perpendicular direction onto the surface of the liquid. d) Schematic picture of a measurement. With the normal force transducer of the ARES G2, ΔF can be measured from 0.001 to 20 N. Thus, the theoretical minimum of the measurable interfacial tension is $\Gamma = \frac{\Delta F}{U} = 25\,\mathrm{mN/m}$ with a perimeter $U = 0.0402\,\mathrm{m}$, see Equation (5.5).

before and after the change in surface tension is due to the post averaging of the data. The convolution with the rectangular function with a length of 51 points generates a smoothed curve. In chapter 9 the surface tension of different brands of beer are measured, as another example for surface tension measurements with the Wilhelmy plate in combination with the ARES G2.

5.3.2 Interfacial tension

The interfacial tension is the more challenging subject. The two possibilities of moving the plate upwards and downwards are also both applicable. Moving the plate from phase 2, Figure 5.3b, downwards into phase 1, a sharp decrease of the axial force is measured, Figure 5.6b and d. But the sharp decrease only appears within the second run, which means, that the plate is on the one hand completely covered with phase 1 and on the other hand it needs to be coated with the liquid, in which it will immerge (phase 2). The reason might be, that the high viscous samples do not provide a fast enough wetting with a contact angle close to zero, if phase 2 has not coated the plate before. The coating with phase 2

Table 5.1: Calculation of the surface tension Γ_{meas} with Equation (5.5), where the Wilhelmy plate of platin iridium alloy has a perimenter of 40.2 mm. The reference values are taken from [Atkins 07] for water and from the safety data sheet from Wacker in case of PDMS. The downward movement of the plate results in closer values of the surface tension with respect to the reference values Γ_{ref}.

sample	plate movement	ΔF (N)	Γ_{ref} (mN/m) at 25 °C	Γ_{meas} (mN/m) at 25 °C
water	down	$2.85 \cdot 10^{-3}$	72	69.4
water	up	$2.99 \cdot 10^{-3}$	72	73
PDMS	down	$7.55 \cdot 10^{-4}$	21	18.4
PDMS	up	$8.48 \cdot 10^{-4}$	21	20.7

in a first run is only required in the bottom of the plate, which will immerse in phase 2 also in the second run. In the upward movement, phase 1 is inserted into the cup and the Wilhelmy plate is positioned in a way that its bottom tips into this phase. Afterwards the liquid with lower density, phase 2, is added carefully, so that the two immiscible phases do not mix and phase 2 completely covers the Wilhelmy plate. At this stage an excess of phase 2 has to be added, that the Wilhelmy plate is covered until the film of phase 1 drafts from the bottom of the Wilhelmy plate during the upward movement.

Interfacial tension measurements were applied to two immiscible phases. The water phase consists of 50 wt% polyethylene glycol (PEG) dissolved in water and an oil phase consisting of miglyol 812. Miglyol 812 is a Capric/caprylic triglyceride, a combined triester of a blend of capric and caprylic acids [Sasol Germany GmbH 11]. Capric acid is a saturated C_8 fatty acid with the systematic name decanoic acid and caprilyc acid is the corresponding C_8 fatty acid, also called octanoic acid, see Figure 5.5a. The surfactant polyglycerinpolyricinoleate (PGPR) (Figure 5.5b) consists of an oligomer of glycerin which reacts with ricinoleic acid (Figure 5.5c) to an ester. The surfactant PGPR is added to vary the interfacial tension.

In Figure 5.6a and c the plate is moved upwards and in b and d downwards. From the changes in axial force the interfacial tensions are calculated and compared with spinning drop measurements, see Table 5.2.

It could be shown that both measurements, surface and interfacial tension, can be measured

Table 5.2: Calculation of the interfacial tension with Equation (5.5). The reference values, Γ_{ref}, were measured by spinning drop. The samples consist of a water phase of polyethylene glycol (PEG) 50 wt% dissolved in water and an oil phase of miglyol 812, see Figure 5.5a. The surfactant polyglycerinpolyricinoleate (PGPR) is added to vary the interfacial tension, see Figure 5.5b and c. The nomenclature refers to the addition of PGPR. The comparison of the interfacial tension Γ_{meas} values with the reference values Γ_{trg} show that the downward movement of the plate, especially for the sample with surfactant, achieves closer value with respect to Γ_{ref}.

sample	plate movement	ΔF (N)	Γ_{meas} (mN/m) at 25 °C	Γ_{ref} (mN/m) at 25 °C
without	down	$3.06 \cdot 10^{-4}$	7.6	8.7
without	up	$3.97 \cdot 10^{-4}$	9.9	8.7
with	down	$2.35 \cdot 10^{-4}$	5.83	5.5
with	up	$2.99 \cdot 10^{-4}$	7.4	5.5

with the Wilhelmy plate in combination with the ARES G2. The theoretical calculated minimum interfacial tension of 25 mN/m was not the limit for the measurements. Interfacial tensions down to 5.8 mN/m were measured. Within the experimental procedure the Wilhelmy plate is either moved downwards or upwards. The measured surface tension values for water and PDMS are closer to the reference values with the upward movement. In contrast to this the interfacial tension measurements of the two immiscible fluids (PEG dissolved in water and miglyol oil) either without surfactant PGPR or with PGPR yield a quantitative better agreement with the reference values for the downward movement of the plate.

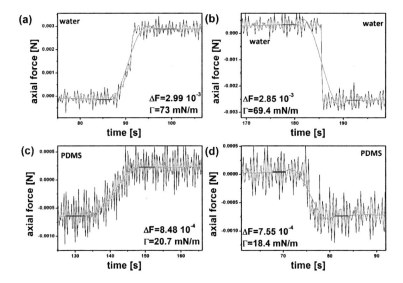

Figure 5.4: Measurement of the surface tension of water and PDMS with $M_n = 5\,\mathrm{kg/mol}$ by upward and downward movement of the Wilhelmy plate. Upward (a) and downward movement (b) of the Wilhelmy plate into distilled water results in a jump of the axial force. No significant difference is observed between the two different measurement techniques, applied at $T = 25\,^\circ$. In (c) and (d) the axial force is measured from a commercial silicon oil. The calculated surface tensions via Equation (5.5) are listed in Table 5.1 together with the literature values for water, [Atkins 07], and for PDMS from the safety data sheet of Wacker. The moving of the plate was set to $0.05\,\mathrm{mm/s}$. The yellow bars sign the region, which are used to determine the mean values right before and after the jump of the axial force.

(a)

C_{10}

C_8

(b)

RO

OR

RO

O

OR

n=1-4

R=H and /or ricinoleic acid
and/or polyricinoleic acid

(c)

H₃C

H

OH

OH

Figure 5.5: a) Structure of capric (C_{10}) and caprylic acid (C_8) the saturated fatty acids of the combined triester miglyol 812 and b) of polyglycerinpolyricinoleate (PGPR) with ricinoleic acid displayed in c) [Belitz 08].

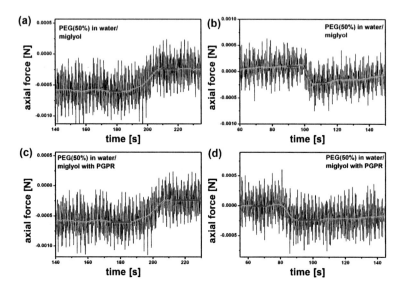

Figure 5.6: a) and b) Interfacial tension measurements of two immiscible fluids. The phase with the higher density is polyethylenglycol (PEG) dissolved in water (50 wt%) and is measured against the oil miglyol a triester of capric and caprylic fatty acids, see Figure 5.5a. c) and d) The surfactant polyglycerinpolyricinoleate (PGPR) is added to miglyol 812, see Figure 5.5b and c. The upward movement of the Wilhelmy plate, (a) and (c), result in less sharp changes of the axial force compared to the downward movement in (b) and (d). The measurement temperature was $T = 25°$. The moving of the plate was set to 0.05 mm/s for both directions. The yellow bars sign the region, which are used to determine the mean value right before and after the jump of the axial force. The interfacial tensions are summarized in Table 5.2 together with reference values.

6. Colloidal dispersion

Dispersions with a dispersed phase of a compact polymer in a liquid are called polymer colloid. Within this thesis it is assumed, that the polymer colloids behave like a solid. In the introduction the influence of the volume fraction on the flow properties of polymer colloids is shown in Figure 1.4. If the volume fraction Φ_{Vol} exceeds a critical value of approximately 0.58, a glassy behavior of the dispersion is observed in a hard sphere model, Figure 1.4. Dispersions, especially suspensions with a solid dispersed phase, are important in numerous applications as shown by Pal et al. [Pal 06] and references cited there. Some dispersions shall be mentioned as an example: recording media which are manufactured using a suspension of magnetic particles; paints with the content of fine insoluble solid particles to achieve the desired color; cosmetics and toiletries like sunscreen lotions and toothpaste, respectively.

6.1 Definition of colloidal glass

Within literature different types of glasses are described [Kob 02, van Megen 94, Götze 92]. A main difference is made between strong and fragile glasses, see Figure 6.1 [Kob 02, van Megen 94, Angell 94]. Silica, SiO_2, is an example for a strong glass. Strong glasses possess activated processes with a constant energy barrier. With a decreasing temperature or inverse density the thermodynamic properties like pressure, specific heat, enthalpy show a second order phase transition, due to the freezing of the configurational degrees of freedom. But phonon activated hopping mechanisms in the glassy state can create a

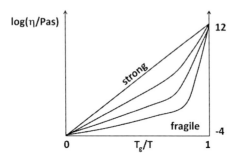

Figure 6.1: Viscosity of strong and fragile glass-forming liquids as a function of T_g/T, where T_g is the glass transition temperature at which the viscosity value is arbitrarily chosen as 10^{12} Pas [Angell 94].

restoring of the ergodicity, which cannot be found in fragile glass. Ergodicity describes the average characteristics of a system, which develops as a function of time. An important characteristic of ergodic systems is the same result of the two possibilities averaging the properties of the systems. Either the time average of one property is calculated or the average of different properties over the entire space [van Megen 94, Brader 09].

Considering the temperature effect on the rheological properties strong and fragile glasses behave differently. In strong glasses the temperature dependent viscosity shows Arrhenius behavior, whereas the fragile glasses posses a stronger temperature dependence, see Figure 6.1. The glass transition in the mode coupling theory for temperature dependent relaxation processes is defined as the temperature where the relaxation time becomes comparable with the experimental time in dimension of seconds. It appears at a critical temperature T_c which is approximately 20% above a temperature T_g in Kelvin, where T_g is the temperature, first proposed by Laughlin and Uhlman, at which the viscosity value is arbitrarily chosen as 10^{12} Pas, see Figure 6.1 [Angell 94, Götze 92]. The behavior of temperature dependent fragile glasses is expressed by the Vogel-Fulcher-Equation [Fulcher 25, Vogel 21]:

$$\eta = \eta_0 \exp\left(\frac{E_a}{R(T - T_{\mathrm{VF}})}\right) \tag{6.1}$$

with E_a as energy barrier and T_{VF} as Vogel-Fulcher temperature. The different temperatures can be ordered as follows: $T_{\mathrm{VF}} \ll T_g \ll T_c \ll T_m$ with T_m the melting temperature.

Polymers are an example for a temperature dependent fragile glasses. Polymers have the characteristic that they are able to rotate and reptate along their chains [Dealy 06]. Close to T_g these orientation processes are frozen and the chains crystallize with an amorphous structure. Experimental investigations of the structural relaxation of polymers in the glassy region are made for example by multidimensional NMR techniques as done by Spiess et al. [Heuer 95, Tracht 98].

Within this thesis the glass transition of the colloidal dispersion is based on the variation of the volume fraction. The temperature is assumed to have no influence on the relaxation processes as visualized in the Angell plot Figure 6.1, but is merely used to adjust the volume fraction as described below in Section 6.4. The volume fraction of the dispersed phase subdivides a colloidal dispersion into groups with different physical properties, Figure 1.4. Above a volume fraction of 0.58 the glassy state is reached. With an increasing volume fraction, the particles are more and more trapped in persistent cages of neighbor particles (='cage'-effect). Some characteristics about the glass transition, either induced by temperature or volume fraction variation, are listed in the following for glass-forming systems [Kob 02]:

- disorder and/or frustration in the structure or in the interactions

- no long range order

- non-ergodic phase transition

- time dependent correlation function shows a stretched exponential α-relaxation at long times and β-process with a power law behavior which merges into a plateau related to the cage effect.

In Figure 6.2 an example of a time dependent correlation function at low and high volume fraction, i.e. in the fluidic and glassy region, respectively, is shown as it can be found in [Kob 02]. At low volume fractions the fluidic state is approximated with only one relaxation called α relaxation. At short times exists the ballistic regime with a square time dependence of $\Phi(t)$ based on a Taylor expansion of the equation of motion [Kob 02]. For long time scales, after the microscopic regime, the Debye relaxation is described by an exponential function of the density-density autocorrelation function. Whereas at high volume fractions two distinct relaxation processes occur. At short times there is again a ballistic regime, which runs into the microscopic regime with a plateau. This time window is called the β relaxation. This plateau appears due to the 'cage'-effect mentioned above.

The β relaxation describes the freezing of the particles motion. Only at a long time scale the particles are able to leave their neighbor cage and thus the correlation function decays to zero in a stretched exponential α relaxation process. Hence the α relaxation describes the break up of the cages.

This work will focus on the rheological characterization of polymer colloids. It is assumed that the polymer dispersed phase behaves in a first approximation like solid spheres. The polymer colloid is therefore also called colloidal suspension. The investigated system exhibits the advantage of a simple constitution of assumed hard spheres in a structureless, incompressible liquid. Additionally there is no phonon activated transport which could restore the ergodicity. Crystallization is suppressed by a polydisperse distribution with 17 % polydispersity of the dispersed particles. Here the polydispersity is the standard deviation of a Gaussian distribution [Bronstein 08]. As a last point it should be mentioned, that in colloidal suspensions hopping processes are negligible, which could qualitatively change the intermediate time window [Fuchs 95]. All in all hard sphere suspensions are an ideal system to describe the impact of the cage effect on the relaxation process via mode coupling theory, see Section 6.2.

6.2 Mode coupling theory

The mode coupling theory (MCT) is an approach to describe the relaxation processes beyond the microscopic regime for suspensions at the critical temperature T_c. The MCT uses equation of motions to describe the dynamics in a glassy system, i.e. the density fluctuation functions [Fuchs 95]. At rest the particles are surrounded by neighbor particles which form a cage with increasing density. If the glassy state is reached, each particle is trapped in a cage of neighbor particles and the movement of one particle needs the rearrangement of the cages cooperatively. Thus the dynamics of the system are slowed down, which is described by the density fluctuations $\delta\rho(\mathbf{q}, t)$. The autocorrelation of the density fluctuations $F(\mathbf{q}, t)$ is the static structure factor $S_{\mathbf{q}}$ at time $t = 0$ with $\mathbf{q}$ being the scattering vector. To eliminate the static structure factor a density-density autocorrelation relaxation is introduced as $\Phi_{\mathbf{q}}(t) = F(\mathbf{q}, t)/S_{\mathbf{q}}$ [Götze 92]. Due to Fourier Transformation, Section 2.1, the scattering vector is correlated to the real dimensions by $\mathbf{q} \approx 2\pi/d$ (Equation (8.18)), where d is the average particle distance [Fuchs 95]. The typical density correlation function Φ can be described by the MCT with a differential

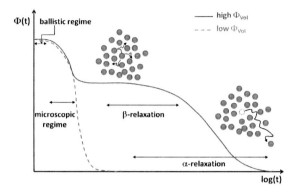

Figure 6.2: Time dependent density-density autocorrelation function at low and high volume fractions Φ, which corresponds to a fluidic and glassy state, respectively [Kob 02]. At low volume fractions the liquid state shows only one relaxation process, the α relaxation. Whereas the glassy system reflected by the high volume fraction curve, possess a β relaxation, effected by the trapped particles in a neighbor cage, beside an α relaxation. It should be mentioned, that the α relaxation of liquids is a Debye exponential decay, whereas in a glass it is a stretched exponential dependency.

equation of a damped harmonic oscillator and an additional integral containing the memory function $m_{\underline{q}}(t - t')$:

$$\ddot{\Phi}(t)_{\underline{q}} + \nu\dot{\Phi}_{\underline{q}} + \Gamma_{\mathbf{q},\text{MCT}}(t)\left[\Phi_{\underline{q}}(t) + \int_0^{t'} dt'm_{\underline{q}}(t - t')\dot{\Phi}_{\underline{q}}(t')\right] = 0 \qquad (6.2)$$

where $\Gamma_{\mathbf{q},\text{MCT}} = \Omega_{\mathbf{q}}^2/\nu$ is the decay rate of $\Phi_{\underline{q}}(t) \to \exp(-\Gamma_{\mathbf{q},\text{MCT}}t)$ with $\nu = \sqrt{k_B T/m}$ as thermal velocity, $\Omega_{\underline{q}}^2 = \nu^2\underline{q}^2/S_{\underline{q}}$ the characteristic frequency of the liquids dynamic and $S_{\underline{q}}$ is the wave vector dependent structure factor. The structure factor mimics the microscopic dynamics determined by density fluctuations. The initial conditions of the differential equation of the density autocorrelator are:

$$\Phi(t = 0) = 1$$
$$\frac{\partial \Phi}{\partial t} = 0 . \qquad (6.3)$$

The inertia term $\ddot{\Phi}(t)$ is related to the acceleration of the colloids. Investigation of polymer collloids in a liquid dispersion medium causes collisions between colloids and solvent molecules. These collisions reduce the acceleration of the colloidal particles to a minimum and, thus, the inertia term is negligible on a Brownian motion time scale, where the damping terms dominate.

Within this thesis the greek letter "Γ" is used in the different chapters for several parameters (interfacial and surface tension, decay rate), therefore, the decay rate Γ_{MCT} has the index "MCT" within this thesis.

An exact expression for the memory function $m_{\underline{q}}(t - t')$ with t' smaller than t is not known. It is assumed to be a linear combination of all slow variables or modes, which explains the name of the theory: mode coupling theory [Götze 91, Götze 92, van Megen 94]. The memory function can be approximated by projecting the fluctuating forces onto density pairs or modes $\underline{k}$ and $\underline{p}$ and factorizing the resulting pair-density correlation function as:

$$m_{\underline{q}}(t) = \sum_{\underline{k}+\underline{p}=\underline{q}} V(\underline{q}, \underline{kp}) \Phi_{\underline{k}}(t) \Phi_{\underline{p}}(t) . \tag{6.4}$$

The coupling vertex $V(\underline{q}, \underline{kp})$ combines the various coefficients of the polynomials of the memory function in one vector.

6.3 Schematic $F_{12}^{(\dot{\gamma})}$ model

The MCT approach is based on the consideration of all wave vectors $\underline{q}$. A simplification of this method is the reduction of the dependency on all wave vectors $\underline{q}$ to a representative one. The spatial fluctuation is therefore no longer respected, whereas the time density fluctuation at a single structure $S(\underline{q})$ is observed. Thus, Equation (6.2) simplifies in the schematic $F_{12}^{(\dot{\gamma})}$ model to:

$$\dot{\Phi} + \Gamma_{MCT}(t) \left[\Phi(t) + \int_0^{t'} dt' m(t - t') \dot{\Phi}(t') \right] = 0 . \tag{6.5}$$

The memory function is approximated with the first two polynomial of the correlator Φ or Equation 6.4, respectively. This, together with the single wave vector dependency, leads to the name *schematic $F_{12}^{(\dot{\gamma})}$ model*, where $\dot{\gamma}$ is used to declare the application of the MCT based theory under shear. The memory function simplifies to [Fischer 91]:

$$m(t) = v_1 \Phi(t) + v_2 \Phi^2(t) \tag{6.6}$$

with v_1 and v_2 as static structure factors. The static structure factors deal as coupling parameters and describe the increasing cage effect with increasing values of $v_1, v_2 \geqslant 0$.

The schematic $F_{12}^{(\dot{\gamma})}$ model of the quiescent state was extended by Professor Fuchs (University Konstanz, Germany) and Professor Brader (University Fribourg, Switzerland) to oscillatory shear. Under shear the model has to describe two competitive phenomena: increasing particle interactions leading to the non-ergodicity transition and the opposed shear-induced decorrelation. It is worth to mention that flow-induced ordering is assumed to be negligible in the range of applied Peclet numbers $Pe_0 \ll 1$. The Peclet number defines the ratio of shear to particle dynamics determined by Brownian motion:

$$Pe_0 = \dot{\gamma} R_h^2 / D_0 \qquad (6.7)$$

where D_0 is the diffusion coefficient of particles with hydrodynamic radius R_h, [Fuchs 03]. An estimation for the here proposed Pe_0 is in the dimension of 10^{-5} with particles with a hydrodynamic radius of approximately $R_h = 100\,\mathrm{nm}$, shear rates of $\dot{\gamma} = 1\,\mathrm{s}^{-1}$ and a diffusion coefficient $D_0(\mathrm{H_2O}) = 2.3 \cdot 10^{-9}\,\mathrm{m}^2/\mathrm{s}$ taken from water for a first estimation. Pe_0 describes the influence of shear to the single particle. In contrast to this the Weissenberg number, also called dressed Peclet number, expresses the influence of shear to the structural relaxation time τ:

$$Pe = \dot{\gamma}\tau \; . \qquad (6.8)$$

In the region of $Pe > 1 \gg Pe_0$ the nonlinear rheology is dominated by the competition of the contrary phenomena, structural rearrangement and influence of shear. Figure 6.3 shows the shear-induced decorrelation of the memory, which is not caused by shear-induced ordering [Fuchs 03, Crassous 08a]. Brownian motion plays an important role by smearing out the bars indicated by the circles in Figure 6.3b and causes the decay of λ_y for large times, which means a decay of the density correlator. Under nonlinear rheology advection enforces the decorrelation processes against the mechanism of the cage effect. Within the memory function the shear melting effect of glass is added to Equation (6.6) and results in:

$$m(t) = \frac{v_1 \Phi(t) + v_2 \Phi^2(t)}{1 + (\dot{\gamma} t / \gamma_c)^2} \qquad (6.9)$$

with γ_c as deformation needed to break the cages.

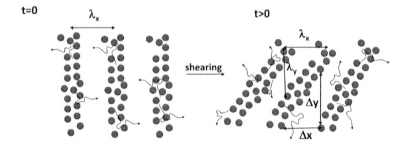

Figure 6.3: The circles show the particles in cages and their density fluctuations with an initial distance λ_x at $t = 0$. Due to advection the density fluctuations are moved correspondingly to the applied shear in x direction, which is reflected by the picture at $t > 0$. Additionally with an applied shear, a second wavelength λ_y appears. The dependence of λ_y on the shear is described by $\lambda_x/\lambda_y = \Delta x/\Delta y = \dot\gamma t$, [Fuchs 03, Crassous 08a]. Brownian motion causes a smearing out into $x-$ and $y-$direction which causes an accelerated decorrelation in contrast to the relaxation only in $x-$direction at $t = 0$. This makes clear that already small Peclet numbers induces decorrelation of the memory but without flow-induced ordering.

6.3.1 Oscillatory rheology

As a general description of the time-dependent shear stress as function of shear rate $\dot\gamma$ is given by a Green-Kubo relation [Green 54, Kubo 57, Brader 07, Brader 08]:

$$\sigma(t) = \int_{-\infty}^{t} dt' \dot\gamma(t') G(t, t')$$ (6.10)

where $G(t, t')$ is the shear modulus nonlinear dependent on $\dot\gamma$. Under oscillatory shear the steady shear rate $\dot\gamma$ is replaced by the time dependent shear rate $\dot\gamma = \gamma_0\omega_1\cos(\omega_1 t)$ (time derivative of the sinusoidal strain amplitude $\gamma = \gamma_0\sin(\omega_1 t)$), where the amplitude $\gamma_0\omega_1$ is main important for characterizing the stress response. The memory function in Equation (6.9) is adjusted for oscillatory shear by exchanging the shear rate against the overall amplitude of applied strain and frequency:

$$m(t) = \frac{v_1\Phi(t) + v_2\Phi^2(t)}{1 + (\frac{\gamma_0\omega_1 t}{\gamma_c})^2} .$$ (6.11)

It implies that each amplitude γ_0 can melt a glass, even if in a quiescent state the system would not show a decorrelation behavior. Accordingly it is unclear whether a linear regime can be defined or not [Brader 10]. A closed theory for the shear stress in the time domain under large amplitude oscillatory shear is given by the sum of Equation (6.10), Equation (6.11), the equation of motion, Equation (6.5), and the definition for the shear modulus $G(t, t')$:

$$G(t, t') = v_\sigma \Phi^2(t) + \eta_\infty \delta(t - t') \tag{6.12}$$

where the second term on the right hand side considers hydrodynamic interactions for upcoming viscous processes on time scales much shorter than the structural relaxation time. These are processes which do not require any structural relaxation [Crassous 08a]. Additionally $\delta(t)$ is the delta function and η_∞ is the high frequency viscosity. In [Crassous 08a] the relation between high shear and high frequency viscosity was found as:

$$\lim_{\omega_1 \to \infty} G''/\omega_1 = \eta_\infty^{\omega_1} = \eta_\infty \tag{6.13}$$

and

$$\eta_\infty^{\dot{\gamma}} = \lim_{\dot{\gamma} \to \infty} \sigma(\dot{\gamma})/\dot{\gamma} = \eta_\infty + \frac{v_\sigma}{2\Gamma_{\mathrm{MCT}}} \tag{6.14}$$

based on the assumption, that the memory function $m(t)$, Equation (6.11), vanishes for high shear and that Equation (6.5) yields an exponential decay of the correlator, $\Phi(t) \to \exp(-\Gamma_{\mathrm{MCT}} t)$ for $\dot{\gamma} \to \infty$.

6.4 Experiments

Dense colloidal suspensions were simulated to predict the nonlinear rheological behavior under large amplitude oscillatory shear (LAOS) with a simplified MCT model, the schematic $F_{12}^{\dot{\gamma}}$. The project merges the work of three different parties. At the Universtity Bayreuth (Germany) in the group of Professor Ballauff (now Helmholtz Gemeinschaft Berlin, Germany) the model systems were synthesized [Siebenbürger 09], which were measured with FT-Rheology at the Karlsruhe Institute of Technology in the group of Professor Wilhelm, where the invention of FT-Rheology is manifested [Wilhelm 02]. The theoretical predictions were simulated at the University Konstanz by Fuchs and Brader (now University Fribourg, Switzerland) et al. [Brader 10].

6.4.1 Characterization of the colloidal suspension

The characteristics of the latices investigated by FT-Rheology are taken from Siebenbürger et al. [Siebenbürger 09], because of their synthesis and characterization in in the group

of Professor Ballauff. As mentioned in Section 6.1 crystallinity is avoided by a polydisperse distribution. The polydispersity of the investigated system is 17 % determined by
the standard deviation σ_d of a Gaussian distribution, see Equation (7.25). The particles
consist of a solid core of poly(styrene) with a chemically grown shell of a thermosensitive network of crosslinked poly(N-isopropylacrylamide) (PNIPAAM) [Siebenbürger 09].
The network was developed by N, N'-methylenebisacrylamide and yielded a crosslinking
of 2.5 mol%. The chemical structures of the contributing molecules is shown in Figure
6.4. The thermosensitive shell causes a temperature dependent hydrodynamic radius of

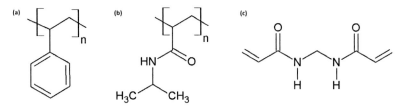

Figure 6.4: a) Chemical structure of the polystyrene core b) of the shell of poly(N-isopropylacrylamide),
which was crosslinked to a network with c) N, N'-methylenebisacrylamide.

$R_H = -0.7796T + 102.4096$ expressed in nanometers with T as the temperature in °C and
below 25 °C [Siebenbürger 10]. Cryogenic transmission electron microscopy (cryo-TEM)
and dynamic light scattering (DLS) were used to investigate the structure and swelling of
the particles [Crassous 06, Crassous 08b]. The swelling of the particles is related to the
swelling of the polymer network building the shell, see Figure 6.5. The transition to a

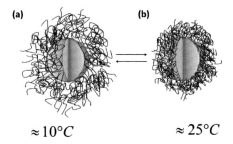

$$\approx 10°C \qquad\qquad \approx 25°C$$

Figure 6.5: The hydrodynamic radius R_H of the core-shell particles depends on the temperature. With
increasing temperature R_H decreases. Above a temperature of 32 °C the polymer network of
PNIPAAM collapses [Siebenbürger 10].

dynamic arrested structure for monodisperse hard spheres is at a volume fraction Φ_{Vol} of 0.58, as shown in Figure 1.4. For the here investigated polydisperse colloidal suspension, the glass transition is shifted to a higher volume fraction of $\Phi_{Vol} = 0.64$. The random close packing volume fraction shifted from $\Phi_{Vol} = 0.64$ for monodisperse systems to $\Phi_{Vol} = 0.68$ for the here investigated system. The increase of the volume fraction for the glass transition as well as for the random close packing can have several reasons. The first one is the 17 % polydispersity of the dispersion. As second reason, the assumption, that the polymer colloids behave like hard spheres also at high volume fractions, could not be fulfilled. Additionally the dynamic light scattering measurements include a deviation from the true particle diameter.

The synthesis with an anionic initiator of the polymerization remains some charges at the surface of the shell. Therefore all experiments were done in an aqueous solution of 0.05 mol/l KCl to shield the charges. Thus, only steric stabilization takes place [M. Siebenbürger 06]. Additionally the solid content of the suspension was determined by comparing the weight before and after drying and resulted in 8.35 wt%. The effective volume fraction Φ_{eff} was calculated by using the correlation between mass concentration c (g/mL), hydrodynamic radius R_H (m^3) and the correlation found in Figure 6 of [Siebenbürger 09], given by $c \cdot R_H = 9.67 \cdot 10^{-17} \, g \cdot \Phi_{eff}$. The different temperatures investigated herein yield different effective volume fractions of $\Phi_{eff} = 0.57$ at 20 °C, $\Phi_{eff} = 0.60$ at 18 °C and $\Phi_{eff} = 0.65$ at 15 °C. Thus, measurement below and closely above the glass transitions are made.

6.4.2 FT-rheological experiments

The measurements were applied on a strain controlled ARES G1 (100FRT) rheometer (Rheometrics Scientific) with a cone plate geometry of 50 mm diameter and an angle of 0.04 rad. Further specifications can be found in Section 11.2.1. The measurement conditions are kept constant by a water bath device, a solvent trap with a sponge drawn with water and by an additional thin film of paraffin around the borders of the sample in the geometry. The latter two provisions were used to prevent evaporation. The FT-rheological measurements were performed at 15.1, 18.4 and 20.9 °C in contrast to the linear measurements which were applied at 15, 18 and 20 °C. In the group of Professor Ballauff (Helmholtz-Gemeinschaft Berlin, Germany) several linear measurements were applied on an MCR 301, whereas the FT-rheological experiments were measured in Karlsruhe at the ARES G1 as part of this thesis. The change of the rheometers (MCR 301 for linear measurements to ARES G1 for the nonlinear measurements) and the change in location results in different ambient conditions. The temperatures 15.1, 18.4 and 20.9 °C used for the nonlinear measurments fit to the Φ_{eff}=0.57, 0.60 and 0.65 calculated above for temperatures 15, 18 and 20 °C used for the linear measurements (Section 6.4). The differences in temperature were adjusted from comparative measurements in the linear regime between the ARES G1 and the stress controlled rheometer MCR 301 (Anton Paar). Before the oscillatory time sweep measurements were measured a preshear with steady shear rate $\dot{\gamma}$ of $100\,s^{-1}$ for 200 s and a soak time of 10 s is applied. Several cycles were detected to secure oscillatory stationary state behavior. For the excitation frequency of 1 Hz time sweep tests with 40 oscillations are applied, for 0.1 Hz 10 oscillations and for 0.01 Hz 9 oscillations. The post processing of the raw data performs a discrete, complex, half-sided fast Fourier transformation as described in Section 2.1 and [Wilhelm 02, Wilhelm 99, Bracewell 86].

6.5 Results

The FT-rheological characterization of the polymer colloid was done in a collaboration of three parties. The synthesis was done at the University Bayreuth (Germany) in the group of Professor Ballauff (now Helmholtz Gemeinschaft Berlin, Germany) as well as the linear rheological measurements. The thereoretical calculations and simulations of the nonlinear behavior were done at the University Konstanz (Germany) in the group of Professor Fuchs. The FT-rheological measurements are part of this PhD thesis. The expertise of nonlinear rheology with soft matter was used to characterize the large amplitude oscillatory shear behavior of the colloidal suspensions. Without the experimental results the simulated behavior could not be evaluated. Thus, the FT-rheological measurements are necessary to evaluate the extending of the MCT based theory into the nonlinear regime.

The theoretical predictions based on the MCT model are compared with experiments of the colloidal suspension at three different temperatures to include different volume fractions from nearly fluid to a glassy system. Two experiments in the linear regime, the measurement of the flow curve and the linear moduli G' and G'' as a function of frequency, were used to determine the simulation parameters for the nonlinear prediction with the schematic $F_{12}^{\dot{\gamma}}$ model, see Figure 6.6. This means the schematic model parameters were not fitted to the experimental large amplitude oscillatory shear data, but were fully determined in advance to predict with linear properties the nonlinear behavior. The qualitative and quantitative agreement will be shown in the following. First a physical meaning of the viscous modulus $G''(\omega)$ should be introduced. It reflects the dissipation of energy over an oscillation cycle in the linear as well as in the nonlinear regime. It derives from the nonlinear shear stress in the time domain, which is still periodic with $2\pi/\omega_1$ even though it is no longer sinusoidal. Therefore is can be expressed in Fourier series:

$$\sigma(t) = \gamma_0 \sum_{n=1}^{\infty} G'_n(\omega)\sin(n\omega_1 t) + \gamma_0 \sum_{n=0}^{\infty} G''_n(\omega)\sin(n\omega_1 t) \tag{6.15}$$

where G'_n and G''_n are frequency dependent Fourier coefficients [Brader 10]. Furthermore the energy dissipated per unit volume of material of one oscillation cycle is:

$$E_d = \int_{-\pi/\omega_1}^{\pi/\omega_1} dt\sigma(t)\dot{\gamma}(t). \tag{6.16}$$

Under oscillatory shear the time dependent strain is defined as $\gamma(t) = \gamma_0\sin(\omega_1 t)$ with its time derivative $\dot{\gamma}$. Substituting the time dependent shear rate $\dot{\gamma}$ and Equation (6.15) in the dissipative energy term, Equation (6.16), leads to:

$$E_d = \gamma_0^2 \pi G''_1(\omega) . \tag{6.17}$$

Hence G''_1 is the dissipative energy over one oscillation cycle in the linear and in the nonlinear regime. The elastic, i.e. reversible storage energy, is described by the remaining coefficients G'_1 and $G''_{n>1}$ [Brader 10].

Accordingly to the notation in other LAOS references, [Wilhelm 02], Equation (6.15) can be rewritten to display a reminiscent form of the Fourier series expressed with intensities I_n and phases δ_n of the harmonics :

$$\sigma(t) = \gamma_0 \sum_{n=1,odd}^{\infty} I_n(n\omega_1)\sin\left[n\omega_1 t + \delta_n(n\omega_1)\right] \tag{6.18}$$

with $I_n = |G'_n + iG''_n|$ and a phase shift defined as $\delta_n(n\omega_1) = \arctan(G''_n/G'_n)$ whereas if $n=1$ it is the fundamental harmonic and with $n>1$ and $n=$ odd that are the higher harmonics.

6.5.1 Linear measurements

The linear measurements were conducted in the group of Professor Ballauff. The theoretical predictions were calculated in the group of Professor Fuchs and Professor Brader. The parameters for the nonlinear simulation are taken from flow curves as well as from linear G' and G'' measurements. The linear measurements were detected with a stress controlled MCR 301 (Anton Paar) in Bayreuth. Figure 6.6 shows the frequency dependent elastic and viscous moduli [Brader 10]. This Figure shows the measurement values (symbols) as well as the theoretical fitted curves (solid lines). The parameters used for the fits with the schematic $F_{12}^{\dot{\gamma}}$ model are listed in Table 6.1 and are used for the prediction of the nonlinear behavior in the Figures 6.7 and 6.8.

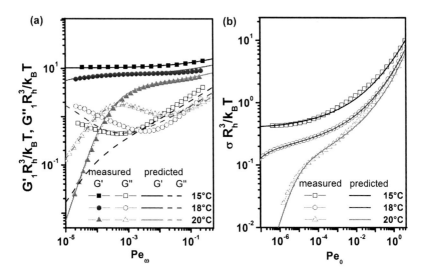

Figure 6.6: (a) Symbols: The experimentally measured linear response moduli for three different temperatures $T = 20.0, 18.0$ and $15.0\,^{\circ}$C with $\Phi_{\text{eff}} = 0.57, 0.60$ and 0.65. Lines: Theoretical Fit with the schematic $F_{12}^{\dot{\gamma}}$ model. (b) Symbols: Flow curves measured at three different temperatures $T = 20.0, 18.0$ and $15.0\,^{\circ}$C, which corresponds to $\Phi_{\text{eff}} = 0.57, 0.60$ and 0.65, respectively. Lines: Theoretical fits with the schematic model parameters. The parameters used for the fits in (a) and (b) are listed in Table 6.1. The parameters for these fitting curves are used for the prediction of the nonlinear behavior in the Figures 6.7 and 6.8 [Brader 10].

The Figure 6.6 possess reduced units defined as:

$$Pe_0 = \frac{\dot{\gamma} R_H^2}{D_0}, \quad Pe_\omega = \frac{\omega R_H^2}{D_0} \tag{6.19}$$

with the diffusion coefficient D_0 determined for the solvent with Stoke's law. The moduli and the shear stress are reduced by:

$$\sigma_{red} = \frac{\sigma R_H^3}{k_B T}, \quad G'_{red} = \frac{G' R_H^3}{k_B T}, \quad G''_{red} = \frac{G'' R_H^3}{k_B T} \, . \tag{6.20}$$

The flow curve measured at $20\,^\circ$C shows a beginning of a Newtonian plateau, which is more and more pronounced if the temperature is decreased and hence the volume fraction increased. At $15\,^\circ$C the flow curve exhibits a total accomplished plateau, which holds also for the lowest Pe_0. The decrease at low Pe_0, after the plateau for $\Phi_{\text{eff}} = 0.60$, suggests an α relaxation time which is out of the experimental time window. The corresponding linear moduli are displayed in Figure 6.6. For the lowest volume fraction a crossover of G' and G'' clearly determines the α relaxation time. With decreasing temperature the glassy state is approached and the crossover point is shifted out of the examined frequency range. But the increase of G'' and a slight decrease of G' for low Pe_ω is an indication for an existing structural α relaxation process. This behavior confirms the flow curve properties. In the glassy system with a volume fraction of $\Phi_{\text{eff}} = 0.65$ the elastic modulus remains constant and the viscous modulus exhibits an increase for small frequencies due to "hopping" initiated relaxation processes. The modeling of the flow curves as well as of the linear moduli at high frequencies show quantitative agreement with the measurements. At low Pe_ω the theoretical fits differ from the experimental behavior due to the MCT based model, which does not include additional relaxation mechanisms like 'hopping' initiated processes.

Specifications about the simulation procedure can be found in [Brader 10]. The results are needed for the determination of the schematic model parameters used in Section 6.5.2. Following parameters are important:

- ϵ: the separation parameter, $\epsilon < 0$ describes the fluid phase and $\epsilon > 0$ the dynamic arrested system in the glassy region

- v_σ: the scaling vertex with unit Pa, see Equation (6.12)

- Γ_{MCT}: the decay rate

- γ_c: the critical strain deformation for break up of the neighbor cages

- η_∞^ω: the high frequency viscosity, defined in Equation (6.13)

Table 6.1 summarizes the model parameters, determined in [Siebenbürger 09], with the corresponding temperatures and volume fractions Φ_{eff}.

Table 6.1: Schematic model parameters determined via linear measurements of the flow curve and the linear moduli G_1' and G_1'' in [Brader 10]. The parameters are used to predict the LAOS behavior with the schematic $F_{12}^{\dot{\gamma}}$ model

T ($^\circ C$)	Φ_{eff}	ϵ	v_σ	Γ_{MCT}	γ_c	η_∞
20.0	0.57	$-2.45 \cdot 10^{-3}$	59	100	0.18	42
18.0	0.60	$-2.2 \cdot 10^{-4}$	85	100	0.19	36
15.0	0.65	$5 \cdot 10^{-5}$	115	105	0.28	24

6.5.2 Nonlinear measurements

The prediction of the nonlinear shear stress in the time domain is done with the schematic model parameters listed in Table 6.1 [Brader 10]. The experimental values for evaluation of the nonlinear theory was done within this thesis. The simulated data are taken with permission from [Brader 10] to show the comparison between theory and experiment. It is worth to mention that the applied schematic model makes a prediction of the nonlinear behavior on the basis of linear measurements and not a simple fitting of the measured data. In Figure 6.7 and 6.8 a direct comparison of theoretical prediction and experiment for two different frequencies, 1 Hz and 0.1 Hz is shown.

Beside the shear stress in the time domain, Lissajous figures, Section 2.2.1, are added to demonstrate the increasing dissipated energy at each oscillation cycle, visualized by an increasing area. For small strain amplitudes a linear response is measured, expressed by a nearly perfect sinusoidal behavior, which becomes more and more distorted with increasing deformation. The full black lines are the MCT data which does not correctly predict the dip or overshoot of the measured data (broken black lines), at the top of the asymmetric peak. The predictions for the lower Pe_ω agree less than in Figure 6.7 due to the more pronounced cut-off shape of the measured shear stress at intermediate strain amplitudes. A common characteristic for both Pe_ω is found in the merging of the minimum and maximum of the measured shear stress $\sigma_{\max}$ and the yield stress σ_y. As soon as $\sigma_{\max}$ is higher than σ_y, indicated as gray lines in the Figures 6.7 and 6.8, a distortion of the shear stress is visible, which remarks the onset of nonlinearity. The shear stress is plotted as a reduced quantity $\sigma_{red} = \frac{\sigma R_H^3}{k_B T}$.

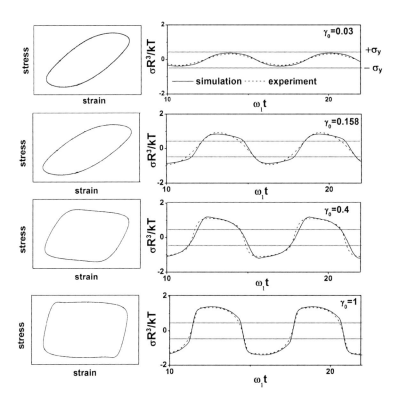

Figure 6.7: Dotted lines are the shear stress in the time domain measured at an excitation frequency of 1 Hz for different strain amplitudes ranging from 0.03 to 5. The temperature was set to $T = 15.1\,^\circ\mathrm{C}$ to measure a glassy system. With increasing strain amplitude an increasing dissipative energy is measured, associated with an increasing area of the closed Lissajous curves. After exceeding the yield stress σ_y, gray lines, the shear stress shows a distorted shape, which reflects the onset of nonlinearity. The MCT based prediction are entered with full lines and give an overall good agreement with the experimental data.

Although the shear stress in the time domain suggests that the schematic model can predict the experimental behavior a quantitative comparison of the nonlinearities $I_{n/1}$ in the frequency spectra is worthwhile. The Fourier Transformation can quantify the nonlinearity, which is difficult by plotting only the distorted shear stress and Lissajous curves, especially for low nonlinearities, see Section 2.2.1. The FT of the shear stress in the time domain results in real and imaginary part, or amplitude and phase spectra, respectively, see Chapter 2. An example of amplitude spectra in the frequency domain are illustrated in Figure 6.10. The spectra exhibit a noticeable difference between experiment, Figure 6.10a, and theory, Figure 6.10c, measured at small strain amplitudes close to the linear regime. The apparent equivalent behavior in the time domain is clearly revised after the FT. The theory predicts clearly visible nonlinear odd harmonics, whereas the experiment shows peaks for even and odd harmonics, but with intensities not far away from noise level. For higher strain amplitudes, Figure 6.10b and d (theory and experiment), show a better agreement, which confirms the similar time behavior, Figure 6.7 and 6.8.

A comparison of the strain amplitude dependent higher harmonics $I_{n/1}(\gamma_0)$ with $n = 3$ and 5 between experiment and simulation with the schematic $F_{12}^{\dot{\gamma}}$ model for a colloidal dispersion in the glassy region ($\Phi_{\text{eff}} = 0.65$) is shown for an excitation frequency of 1 Hz in Figure 6.10. At small strain amplitudes the prediction with the MCT based theory deviates from the nonlinear oscillatory shear measurements. At large deformations the theoretical calculated values show quantitative predictive character of the higher harmonics $I_{n/1}$.

Thus the schematic $F_{12}^{\dot{\gamma}}$ provides a qualitative prediction for small strain amplitudes, which changes into a quantitative agreement for larger strain amplitudes.

The origin of the deviations at small strain amplitudes can be found in the $F_{12}^{\dot{\gamma}}$ model, which shows deviations from the Hookian behavior in this range of deformation. The slow β relaxation is an inherent characteristic of all MCT based models and does not provide an exact prediction for elastic behavior under small deformations.

6.6 Conclusion

The predictive behavior of the MCT based theory is established to extend the linear response regime into the nonlinear regime. The simultaneous determined parameters from the flow curve measurements and the frequency dependent moduli G_1' and G_1'' taken from

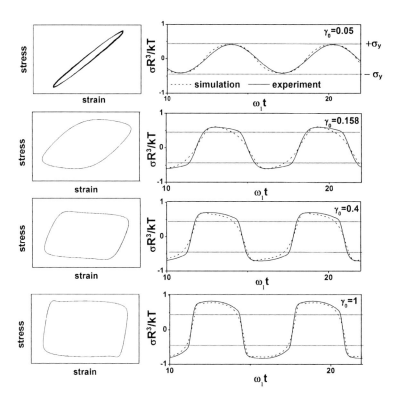

Figure 6.8: Same as in Figure 6.7 but at an excitation frequency of 0.1 Hz. The gray lines indicate the yield stress σ_y, the full black lines are the predictions and the broken black lines are the experimental data.

Brader et al. [Brader 10] yield a quantitative prediction of the nonlinear behavior. This means that linear experiments are able to predict the nonlinear regime. The confirmation of the theory in the nonlinear regime with the here proposed FT-rheological experiments has the advantage that future simulations do not need further evaluation. The unique combination of MCT based simulations and nonlinear oscillatory measurements give an established basis for future calculations of the nonlinear regime. Slight deviations were explored in the prediction of the linear regime with the present schematic $F_{12}^{\dot{\gamma}}$ model. The strain dependence of the higher harmonics in experiment are in dimension of 0.1 % and

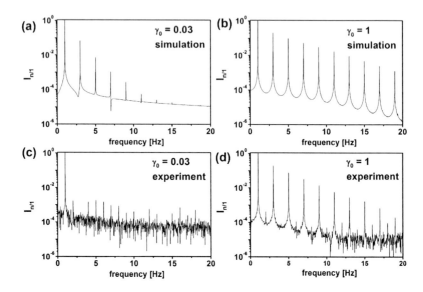

Figure 6.9: Frequency spectra of the oscillatory shear stress in the time domain performed at 1 Hz and 15.1 °C in (a) experiment and (c) theory at $\gamma_0 = 0.03$ and in (d) experiment and (b) theory at an amplitude of $\gamma_0 = 1$.

the theoretical predicted relative third harmonic $I_{3/1}$ has a value of 7 %. The amplified nonlinearity in theory is due to the derivation from the density correlator which possess a slowed relaxation induced by the cage effect. As a second aspect the theory predicts a more symmetrical shape of shear stress in the time domain, whereas the experiment shows a stress overshoot at high strain amplitudes.

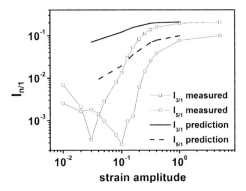

Figure 6.10: Strain amplitude dependent higher harmonics $I_{n/1}(\gamma_0)$ with $n = 3$ and 5 at 15.1 °C from theoretical predictions the schematic $F_{12}^{\dot\gamma}$ model (lines) and measurements (symbols) at an excitation frequency of 1 Hz.

7. Emulsion rheology: Derivation of a nonlinear mechanical master curve for emulsions

Emulsions consist of two immiscible liquid phases, which generates a new material with properties that can vary greatly from the precursors. These products are of strong interest, due to their ability of combining two immiscible fluids as polar and nonpolar substances, like for example water and oil. Emulsions have a wide range of application in daily life, like e.g. in body care products, as wall paints and as food products. The emulsion properties like volume average radius, $\langle R \rangle_{43}$, and polydispersity have an influence on the texture of such systems and their rheological behavior [Larson 99, Pal 06]. Fourier Transform Rheology (FT-Rheology) is used to understand the relation between shear induced structures of dispersions and the non-stationary mechanical nonlinear properties [Hyun 11, Reinheimer 11b].

7.1 The fluid constitutive equation

Within this thesis the properties of emulsions like radius distribution and interfacial tension are of main interest. The combination of simulation and experiment in the large amplitude oscillatory shear allowed the evaluation of this new characterization method. In addition to the experiments under nonlinear shear a comparison with the relevant theoretical predictions in this case constitutive Equation has to be achieved.

To make this comparison, a theoretical model based on a dilute system consisting of buoyancy free droplets of a single Newtonian liquid dispersed in a different, but also Newtonian, matrix is considered. The deformation forces at the interface due to nonlinear oscillatory shear were opposed by the retracting interfacial forces. A higher viscosity of the dispersed phase with respect to the matrix, that means a high viscosity ratio $\lambda = \eta_d/\eta_m$, prevented breakup of the droplets [Grace 82]. As the droplet phase was dilute, coalescence of the droplets was considered negligible and a constant droplet volume distribution was therefore assumed for the simulations and experiments. For the simulation of the shear stress, Batchelor [Batchelor 70] used a linear superposition of the matrix contribution and the contributions originating from each droplet [Grosso 07, Carotenuto 08]:

$$\underline{\underline{\sigma}}(t) = \underbrace{-p\underline{\underline{I}}}_{\text{isotropic term}} + \underbrace{\eta_m \left(\underline{\nabla}\,\underline{v} + \underline{\nabla}\,\underline{v}^T\right)}_{\text{external contribution, linear}} \qquad (7.1)$$

$$\underbrace{-\frac{\eta_m}{V} \int_A (\underline{n}\,\underline{u} + \underline{u}\,\underline{n})\, dA}_{\text{viscous part} \approx 0} \underbrace{-\frac{\Gamma}{V} \int_A \left(\underline{n}\,\underline{n} - \frac{1}{3}\underline{\underline{I}}\right) dA}_{\text{elastic part}} \; .$$

$$\underbrace{\hphantom{-\frac{\eta_m}{V} \int_A (\underline{n}\,\underline{u} + \underline{u}\,\underline{n})\, dA -\frac{\Gamma}{V} \int_A}}_{\text{interfacial contribution, nonlinear}}$$

In Equation (7.1), Γ is the interfacial tension, p is the pressure, $\underline{\nabla}\,\underline{v}$ is the undisturbed velocity gradient tensor and $\underline{\nabla}\,\underline{v}^T$ its transpose, η_m is the viscosity of the continuous phase, V is the volume of the system, $\underline{n}$ is the unit vector orthogonal to the interface between the two phases, $\underline{u}$ is the velocity at the interface and dA is the area of an interfacial element where the integrals are evaluated over the whole interfacial area of the system A. For the prediction of the LAOS behavior of dilute emulsions, only the elastic part of the interfacial contribution was taken into account. The other contributions were either nil, for the Newtonian external force and the isotropic pressure terms, or were considered negligible, such as for the viscous part of the interfacial contribution [Reinheimer 11b]. Accordingly, the shear stress tensor is called the interfacial shear stress $\underline{\underline{\sigma}}_I$.

With the ellipsoidal model of Maffettone and Minale [Maffettone 98, Guido 00], the orthogonal vector $\underline{n}$ to the interface can be calculated via the droplet shape tensor $\underline{\underline{S}}$

[Maffettone 98][Guido 00]:

$$\frac{d\underline{\underline{S}}}{dt} = -f_1(\lambda)\left[\underline{\underline{S}} - \frac{3}{II}\underline{\underline{I}}\right] + f_2(\lambda, Ca(t))$$

$$\cdot (\underline{\underline{D}} \cdot \underline{\underline{S}} + \underline{\underline{S}} \cdot \underline{\underline{D}}) + (\underline{\underline{\Omega}} \cdot \underline{\underline{S}} - \underline{\underline{S}} \cdot \underline{\underline{\Omega}})$$

$$f_1(\lambda) = \frac{40(\lambda + 1)}{(2\lambda + 3)(19\lambda + 16)}$$

$$f_2(\lambda, Ca) = \frac{5}{(2\lambda + 3)} + \frac{3Ca^2}{2 + 6Ca^2}\,. \tag{7.2}$$

With the use of the emulsion time $\tau = \eta_m R/\Gamma$, which is the characteristic relaxation time of the droplet at rest [Carotenuto 08], where R is the radius, η_m the matrix viscosity and Γ the relaxation time, Equation (7.2) was made dimensionless. In Equation (7.2), $\underline{\underline{I}}$ is the second rank unit tensor, $\underline{\underline{D}}$ and $\underline{\underline{\Omega}}$ are the deformation rate and the vorticity tensors, respectively, and II is the second scalar invariant of tensor $\underline{\underline{S}}$, see also [Maffettone 98]. The eigenvalues of $\underline{\underline{S}}$ represent the squared semiaxes of the ellipsoid. The ellipsoid is generally consisting of three different semiaxes as exposed in Figure 1.3 and in literature [Guido 04, Jackson 03] and possess, therefore, no longer rotational symmetry like the spherical shape at rest but two mirror axes. A schematic picture of an ellipsoid is shown in Figure 7.1.

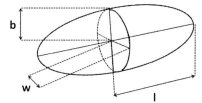

Figure 7.1: Schematic picture of an ellipsoid with two mirror planes. The semiaxes are l, b and w.

Both the diluteness of the sample and its high viscosity ratio $\lambda > 3$ [Grace 82] ensured that the volume distribution remained constant with a droplet shape tensor $\underline{\underline{S}} = \underline{\underline{I}}$ at rest, i.e. for spherical droplets. Quantitative simulations of the nonlinear response of the emulsion via the MM-model were derived for capillary numbers up to slightly higher than unity [Guido 04]. Therefore, the largest radius, R_{max}, experienced the highest capillary number, Ca_{max}, which is defined under LAOS conditions as $Ca_{max} = \frac{\omega_1 \gamma_0 \eta_m R_{max}}{\Gamma}$ with $\dot{\gamma} = \gamma_0 \omega_1$. The experimental input parameters γ_0 and $\omega_1/2\pi$ were adjusted to the radius R_{max} to follow the simulation constraints for comparative experiments.

7.2 Intrinsic nonlinear ratio $^{5/3}Q_0$

The higher harmonic intensities $I_n (n > 1)$ were generally normalized to the fundamental peak I_1 to improve measurement reproducibility. Additionally, the scaling of the normalized intensity of the nth harmonic to the fundamental peak, $\frac{I_n}{I_1} = I_{n/1}$, followed, for small γ_0, a scaling law that included the strain amplitude (Equation (2.11)), see also Figure 7.2 When dealing with Newtonian components in emulsion rheology, the interface deforma-

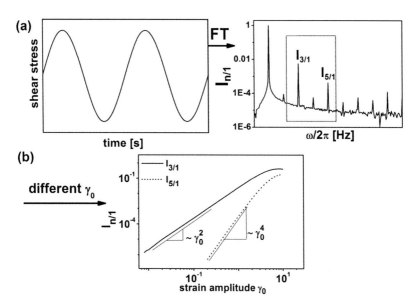

Figure 7.2: (a) Measurement of the shear stress in the time domain of a dilute emulsion PDMS in PIB (see Section 7.7.1) at an excitation frequency of 0.1 Hz and a strain amplitude of $\gamma_0 = 3.135$. The Fourier Transformation yields the frequency dependent spectra with the higher harmonics $I_{n/1}$. (b) Simulation of the normalized $I_{3/1}$ and $I_{5/1}$ as a function of strain amplitude for the MM-B model. The parameters used in the simulation were $\omega_1/2\pi = 0.1$ Hz, $\eta_m = 60$ Pas, $R = 10\,\mu$m, $\Gamma = 3$ mN/m and $\lambda = 3$.

tion under nonlinear oscillatory shear is supposed to be the source of the odd multiples of the fundamental harmonic peak. Therefore, these higher harmonics are related to the droplet size of the included phase and the interfacial tension between the phases in a straightforward manner [Carotenuto 08]. As the fundamental peak I_1 is mainly determined by the Newtonian behavior of the neat Newtonian matrix and dispersed phase, i.e. the viscosity of the two single phases, it was not considered useful in characterizing the interfacial tension or size and distribution of the included phase since it will be dominated by Newtonian contributions and was, therefore, excluded in the following analysis. Of particular interest instead was the intrinsic nonlinear ratio $^{5/3}Q_0$ (calculated using the first two higher harmonics, I_3 and I_5) because it was a useful tool for filtering out the linear part of the experimental signal and included only the nonlinear contributions caused by the deformed interface, Equation (7.3) [Grosso 07]. As a consequence of the scaling law of Equations (2.11) and (2.12), a quadratic dependence on the strain deformation was expected as follows:

$$\lim_{\gamma_0 \to 0} {}^{5/3}Q = {}^{5/3}Q_0 = {}^5Q_0/{}^3Q_0$$

$$= \lim_{\gamma_0 \to 0} \frac{\frac{I_{5/1}}{\gamma_0^4}}{\frac{I_{3/1}}{\gamma_0^2}} = \lim_{\gamma_0 \to 0} \frac{I_{5/3}}{\gamma_0^2} = {}^{5/3}Q_0 \tag{7.3}$$

7.3 Derivation of a nonlinear mechanical master number

The physical parameters that mainly affect the rheological behavior of a dilute Newtonian dispersed phase in a Newtonian matrix are: the viscosities of the matrix and the dispersed phase, interfacial tension, the mean droplet diameter and the droplet size distribution. For the sake of simplicity, only monodisperse blends will be considered in the present study and the first goal of this work is to generate a universal dependency which can relate the dimensionless quantities identified in the model, i.e. the capillary number, $Ca = \frac{\gamma_0 \omega_1 \eta_m R}{\Gamma}$, and the ratio of the dispersed phase and matrix viscosities, $\lambda = \frac{\eta_d}{\eta_m}$. A nonlinear mechanical master curve for emulsions $E(Ca, \lambda)$ would have the advantage that once such a function is established via simulations, it can then be used for every experimental emulsion. Therefore, the amount of data needed to characterize the droplet size or interfacial tension of an emulsion is reduced to a minimum.

It is possible to use FT-rheological analysis of the shear stress to evaluate I_3 and I_5 in the resulting spectra as different physical parameters are varied in experiment and simulation. The scalar quantity $I_{5/3}/\gamma_0^2$ was therefore assumed to be a constant as γ_0 tends to zero. However, of great interest is the fact that this limiting value, which is a constant at small strain amplitudes, is a function of the emulsion properties η_m, R, Γ and the excitation frequency $\omega_1/2\pi$ over a wide range of parameter values.

In starting the derivation of a nonlinear mechanical master number, the viscosity ratio λ is held constant. Figure 7.3 displays a selection of simulated curves showing the intrinsic nonlinear ratio, $I_{5/3}/\gamma_0^2$, for different emulsion characteristics and for $\lambda = 3$. By varying the different emulsion properties and rheological parameters in the following ranges for radii $R = 0.1 - 100\,\mu$m on a logarithmic scale, excitation frequencies $\omega_1/2\pi = 0.1 - 5\,\mathrm{Hz}$ on a linear scale, matrix viscosities $\eta_m = 0.001 - 100\,\mathrm{Pas}$ on a logarithmic scale, viscosity ratios $\lambda = \frac{\eta_d}{\eta_m} = 2.5 - 10$ on a linear scale and interfacial tensions $\Gamma = 2 - 50\,\mathrm{mN/m}$ on a linear scale, the following dependencies were found:

$$\frac{I_{5/3}}{\gamma_0^2} \propto \frac{\omega_1^2 \eta_m^2 R^2}{\Gamma^2} = \frac{Ca^2}{\gamma_0^2} \ . \tag{7.4}$$

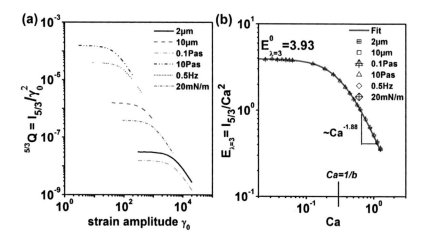

Figure 7.3: a) Simulation of the nonlinear ratio $^{5/3}Q$ as a function of strain amplitude for a constant viscosity ratio of $\lambda = 3$. Each variation in an emulsion property results in a new curve with different plateau value, $^{5/3}Q_0$, for small γ_0. b) Superposition of all curves achieved by plotting E_λ as a function of Ca. The physical parameters used in the simulations are $\omega_1/2\pi = 0.1\,\mathrm{Hz}$, $\eta_m = 60\,\mathrm{Pas}$, $R = 10\,\mu$m, $\Gamma = 3\,\mathrm{mN/m}$ and $\lambda = 3$. The three parameter model, Equation (7.5), quantifies the dependence of E_λ on Ca.

Taking into account the definition of the capillary number, $Ca = \frac{\gamma_0\omega_1\eta_m R}{\Gamma}$, the value for $I_{5/3}$ is clearly seen to be proportional to the square of Ca. A superposition of all the curves shown in Figure 7.3a is achieved by plotting the quantity $E_\lambda = \frac{I_{5/3}}{Ca^2}$ as a function of Ca, which leads to the creation of a single curve, as shown in Figure 7.3b where the simulation conditions are: $\omega_1/2\pi = 0.1$ Hz, $\eta_m = 1$ Pas, $R = 10\,\mu$m and $\Gamma = 10\,$mN/m. Similar to a Carreau model [Larson 99], a three parameter model quantifies the dependence of E_λ on Ca for the shear rate dependent viscosity:

$$E_\lambda = \frac{^0E_\lambda}{1 + (bCa)^c} \ . \tag{7.5}$$

It can be easily demonstrated that the limiting behavior of E_λ is similar to that of $^{5/3}Q_0$ in that $\lim_{Ca\to 0} E_\lambda = E_\lambda^0$. As an example, for $\lambda = 3$, the parameters of Equation (7.5) are estimated by the least square method, where it was found that $E_{\lambda=3}^0 = 3.93$ (plateau value), $b = 2.8$ (representing the onset of decreasing values at $Ca = 1/b = 0.36$) and $c = 1.88$ (describing the slope of the curve at high Ca values). First, it is important to emphasize that E_λ is dimensionless as it is the ratio of two dimensionless quantities. Furthermore, the subscript λ is an important reminder that E_λ depends on the viscosity ratio and, therefore, that different values of E_λ (and E_λ^0) will be observed as λ varies.

7.4 Influence of the viscosity ratio λ

The viscosity ratio $\lambda = \frac{\eta_d}{\eta_m}$ includes the temperature dependence of the viscosity and this causes, therefore, shifts in the plateau values of E_λ^0 with variable λ, see Figure 7.4a, for emulsions under nonlinear shear. The MM-B model is a suitable choice to describe emulsions with viscosity ratios higher than 2.5. At lower values of λ, droplet breakup becomes significant and the MM-B model is less able to describe the experiments. It was found that the plateau value E_λ^0 increases with increasing viscosity ratio. The dependency of the plateau value on the viscosity ratio is modeled in terms of a power law dependency:

$$E_\lambda^0 = E^0 \cdot \lambda^p \ . \tag{7.6}$$

The parameters E^0 and p in Equation (7.6) were estimated by the plateau values of E_λ^0 evaluated from Equation (7.5) at different λ values using a nonlinear regression. With this approach, it was found that $E^0 = 0.64 \pm 0.01$ and $p = 1.63 \pm 0.02$. The excellent agreement between the E_λ^0 values and the power law is shown in Figure 7.4a. In addition, considering that the confidence intervals are very small and that the adjusted R-square statistics (R_a^2) [Draper 98] of the regression was estimated to be $R_q^2 = 0.99885$, this further corroborates the appropriateness of the power law assumption formulated in Equation (7.6). Thus, a

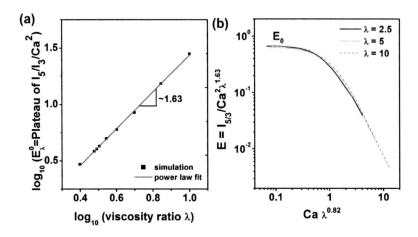

Figure 7.4: a) Computed E_λ^0 values as a function of the viscosity ratio λ (black squares) reported with the power law fit $E_\lambda^0 = E^0 \cdot \lambda^p$ (solid line) with $E^0 = 0.64 \pm 0.01$ and $p = 1.63 \pm 0.02$. b) nonlinear mechanical emulsioncurve $E = I_{5/3}/(Ca^2 \lambda^{1.63})$ as a function of $Ca\lambda^{0.82}$ for different viscosity ratios λ.

new nonlinear, dimensionless coefficient E can be finally introduced and is given by the following relation:

$$E = \frac{E_\lambda}{\lambda^p} = \frac{I_{5/3}}{Ca^2} \cdot \frac{1}{\lambda^p} \ . \tag{7.7}$$

It should be noted that this dimensionless coefficient is a master number as it now includes an explicit dependence on the λ parameter. With Equation (7.7), a superposition of all factors into a single curve is possible by plotting E as a function of $Ca \cdot \lambda^{0.82}$, Figure 7.4b. At low $Ca \cdot \lambda^{0.82}$ (or equivalently γ_0) values, it is possible to estimate the limiting value

E^0 for the E coefficient:

$$\lim_{Ca \to 0} E = \lim_{Ca \to 0} \frac{E_\lambda}{\lambda^p} = \frac{E_\lambda^0}{\lambda^p} = \frac{E^0 \lambda^p}{\lambda^p} = E^0 . \tag{7.8}$$

A crucial point is that this estimated E^0 value, determined via the MM-B model, should be nonlinear mechanically master number for all dilute, monodisperse, emulsions with Newtonian constituents and can therefore be used to determine the physical properties Γ or R. In fact, the following relationship for the dimensional quantities implicitly appearing in Equation (7.8) can be written as:

$$\frac{I_{5/3}}{\gamma_0^2 \omega_1^2} = E^0 \lambda^p \frac{\eta_m^2 R^2}{\Gamma^2} . \tag{7.9}$$

Taking into account the above results of $E^0 = 0.64$ and $p = 1.63$:

$$\lim_{\gamma_0 \to 0} \frac{I_{5/3}}{\gamma_0^2 \omega_1^2} = 0.64 \lambda^{1.63} \frac{\eta_m^2 R^2}{\Gamma^2} . \tag{7.10}$$

Finally, it should be noted that the results obtained with the MM-B model are quantitatively confirmed by a second ellipsoidal model proposed by Yu et al. [Yu 02].

7.4.1 Comparison with the model of Yu and Bousmina

As mentioned previously, the concept introduced here uses ubiquitous quantities, which are not specific to any given model. Therefore, other approaches can also be compared with the results here. For example, the model of Yu et al. uses the Maffettone and Minale model to simulate the droplet shape behavior, but calculates the shear stress with the following Equation [Yu 02] instead of using the Batchelor theory:

$$\underline{\underline{\sigma}} = \frac{2 f_2 K}{II_2} \left(I \underline{\underline{S}} - \underline{\underline{S}} \cdot \underline{\underline{S}} - \frac{2}{3} II \underline{\underline{I}} \right) \tag{7.11}$$

where $\underline{\underline{S}}$ is the non-dimensional droplet shape tensor, $\underline{\underline{I}}$ the unit tensor, I and II the first and second scalar invariant of $\underline{\underline{S}}$, respectively, and $K = \frac{6\Gamma}{5R} \cdot \frac{(\lambda+1)(2\lambda+3)\Phi_{\text{Vol}}}{5(\lambda+1)-(5\lambda+2)\Phi_{\text{Vol}}}$. The dependency of E as a function of Ca with these two different models shows quantitative agreement, as shown in Figure 7.5. Finally, it should be noted that this result supports the assumption made in the calculation of the interfacial shear stress that only the elastic contributions of the interface must be considered, see Equation (7.1), and that any viscous contribution to the interfacial contribution can be neglected in the first approximation using the MM-B model.

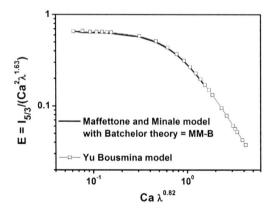

Figure 7.5: Comparison of the nonlinear mechanical emulsioncurve E simulated either with MM-B model (line) and Yu and Bousmina model (symbol).

7.5 Influence of polydispersity

In the previous Section 7.4, the characterization of dilute emulsions via the nonlinear mechanical emulsioncurve E was limited to monodisperse samples. In presence of polydisperse samples, the total stress $\sigma(t)$ of the droplet population, can be calculated as a linear superposition of the contributions pertaining each single droplet:

$$\sigma(t) = \sum_{i=1}^{m} \Psi(R_i)\sigma(R_i)\Delta R_i \ . \tag{7.12}$$

In Equation (7.12) the droplet distribution $\Psi(R)$ has been discretized at m finite equis-paced radii values R_i, where ΔR_i is the difference $R_{i+1} - R_i$. The stress related to each droplet radius can be numerically evaluated through the Equations (7.1) and (7.2). Then, the total stress pertaining the droplet population can be calculated as the sum of the contributions pertaining each droplet size, weighted by the $\Psi(R_i)$ value. Extended sim-ulations for the viscosity ratio dependent E_λ curves were applied for different types of distributions $\Psi(R)$ like the Gaussian distribution, also called normal distribution, and the lognormal distribution, see Section 7.6.1. Figure 7.6 shows an extraction of the applied simulations for two monomodal Gaussian distributions with $\mu = 19.9$ and $29.6\,\mu\mathrm{m}$ with standard deviations $\sigma_d = 1$ and $2\,\mu\mathrm{m}$, respectively. Additionally a bimodal distribution with $\mu_1 = 29.6$ and $\mu_2 = 49.53\,\mu\mathrm{m}$ both with $\sigma_d = 2\,\mu\mathrm{m}$ and a bi-monodisperse distribution with $\mu_1 = 8$ and $\mu_2 = 10\,\mu\mathrm{m}$ were considered. The resulting volume averaged radii and the polydispersities, defined by Equation (7.20), can be found in Figure 7.6. It was found that when the volume averaged radius,

$$\langle R \rangle_{43} = \frac{\sum_{i=1}^{m} R_i^4 \Psi(R_i) \Delta R_i}{\sum_{i=1}^{m} R_i^3 \Psi(R_i) \Delta R_i}, \qquad (7.13)$$

was considered in the capillary number, a master curve was in fact maintained.

The different types of distributions considered in the simulations, where a Gaussian distribution, $\Psi(R) = \frac{1}{\sigma_d \sqrt{2\pi}} \cdot \exp\left(\frac{-(R-\mu)^2}{2\sigma_d^2}\right)$, and a Lognormal distribution, $\Psi(R) = \frac{1}{\sigma_d R \sqrt{2\pi}} \cdot \exp\left(\frac{-(\ln R - \mu)^2}{2(\sigma_d)^2}\right)$, with μ being the mean value and σ_d the standard deviation. Combining the results for p and E^0 from the power law fit together with the volume averaged radius $\langle R \rangle_{43}$, again a nonlinear mechanical master curve for polydisperse systems was now achieved, Figure 7.6. It should be noticed that such universal properties of the emulsion curve have not been appreciated when resorting to other average radii (number average radius, Sauter mean diameter [Schuchmann 05]) and the asymptotic value for the resulting emulsion curve is no longer univocal as $Ca \to 0$. Thus, the volume average radius demonstrates to be the most proper choice to be adopted in the Capillary number. The correlation between nonlinear mechanical emulsion number E^0, intrinsic nonlinear ratio $^{5/3}Q_0$ and emulsion properties is shown with the use of Equation (7.7) in Equation (7.14):

$$\frac{I_{5/3}}{\gamma_0^2 \omega_1^2} = E^0 \lambda^p \frac{\eta_m^2 \langle R \rangle_{43}^2}{\Gamma^2} \qquad (7.14)$$

where $p = 1.63$ and $E^0 = 0.64$, as stated above [Reinheimer 11a]. It should be mentioned that the left side of Equation (7.14) can be evaluated from LAOS experiments using FT-Rheology analysis and includes the intrinsic nonlinear ratio for small γ_0:

$$\frac{^{5/3}Q_0}{\omega_1^2} = 0.64 \lambda^{1.63} \frac{\eta_m^2 \langle R \rangle_{43}^2}{\Gamma^2} \ . \qquad (7.15)$$

Determination of emulsion properties such as volume average droplet radius or interfacial tension for polydisperse emulsions proceeds in a similar way as for monodisperse emulsion characterization. First LAOS experiments were performed at (relatively) small strain amplitudes. Then, the data for I_3 and I_5 were analyzed to detect the range in strain amplitude where power law behavior with respect to γ_0 was seen. With this information the $^3Q_0$ and $^5Q_0$ coefficients as defined in Equation (2.12), were calculated. Finally, the

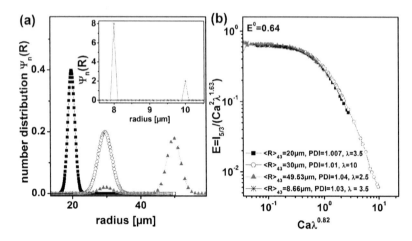

Figure 7.6: (a) Simulations for two monomodal Gaussian distributions with $\mu = 19.9$ and $29.6\,\mu m$ with standard deviations $\sigma_d = 1$ and $2\,\mu m$, respectively, a bimodal distribution with $\mu_1 = 29.6$ and $\mu_2 = 49.53\,\mu m$ both with $\sigma_d = 2\,\mu m$ and a bi-monodisperse distribution with $\mu_1 = 8$ and $\mu_2 = 10\,\mu m$. The resulting volume averaged radii and the polydispersities, defined by Equation (7.20), can be found in the legend in (b). (b) Simulation of the nonlinear mechanical emulsioncurve E for the variety of number distributions displayed in (a) and viscosity ratios $\lambda = 2.5 - 10$. For polydisperse samples, the volume average radius $\langle R \rangle_{43}$ has to be included in the capillary number, Ca, to yield a create a nonlinear mechanical mastercurve comparable to monodisperse conditions. Additionally the viscosity ratio dependence was found by a power law Equation $E_\lambda^0 = E^0 \lambda^p$ with $p = 1.63$ and $E^0 = 0.64$ as reported in [Reinheimer 11a]. The simulation conditions were: $\omega_1/2\pi = 0.1\,Hz$, $\eta_m = 1\,Pas$, $R = 0.1 - 100\,\mu m$, $\Gamma = 5.4\,mN/m$ and $\lambda = 2.5 - 10$.

droplet size $\langle R \rangle_{43}$ was evaluated, if Γ was known, using Equation (7.15) or vice versa if $\langle R \rangle_{43}$ was known the interfacial tension Γ was evaluated. With the simulations a range of $\langle R \rangle_{43}$ between 0.1 to $100\,\mu m$ and for Γ between 2 and $50\,mN/m$ is masked.

7.6 Information about the width of the size distribution

The extension of the simulations from monodisperse to polydisperse samples showed that the higher harmonic ratio $I_{5/3}$ was useful in determining the volume average droplet radius. An interesting feature of concentrated emulsions is that they can produce a very large number of overtones, as many as 289 see Section 3.4, which might be exploited to infer further properties on the droplet population, if they could be measured. For example, as an hypothesis, the ratio of $I_{7/5}$ might be related to another moment of the averaged radius or distribution. The determination of a second, distinct moment to $\langle R \rangle_{43}$, reveals information about the polydispersity of a droplet size distribution. The nonlinear ratio $I_{7/5}$ is chosen, due to the most pronounced intensity after $I_{5/3}$. According to Equation (2.12) a quadratic dependence of $I_{7/5}$ at small strain amplitudes γ_0 is expected, which results in $\frac{^7Q_0}{^5Q_0} := ^{7/5}Q_0$. Following a similar procedure as for $I_{5/3}$, i.e. building the relation between the rheological parameters $^{7/5}Q_0$, ω_1 and the emulsion properties R, η_m, η_d and Γ, a master curve based on $I_{7/5}$ was also achieved once the capillary number and viscosity ratio were included to remove the dependence on individual emulsion properties. However, in contrast to $I_{5/3}$, the definition of the average radius used in the capillary number was here defined as:

$$\langle R \rangle_{54} = \frac{\sum_{i=1}^{m} R_i^5 \Psi(R_i) \Delta R_i}{\sum_{i=1}^{m} R_i^4 \Psi(R_i) \Delta R_i} . \tag{7.16}$$

Again, with other definitions for the average radii, the master number could not be observed if Ca tends to zero. Thus, some information about the droplet dispersion may be then gained from the two different moments of the average radius, $\langle R \rangle_{43}$ and $\langle R \rangle_{54}$. Some analogies between the PDI introduced later in Section 7.7.1 and the new scalar $PDI_{53} = \frac{\langle R \rangle_{54}}{\langle R \rangle_{43}}$ here introduced will be discussed in Section 7.6.1. The nomenclature given above will be transferred to the $I_{7/5}$ analysis with an additional index that refers to the number of the higher harmonics. For example $^{7/5}E_\lambda^0$ refers to the viscosity ratio dependent plateau value related to the master number $^{7/5}E^0$ using the nonlinear power law fit $^{7/5}E_\lambda^0 = ^{7/5}E^0 \cdot \lambda^p$ where $p = 1.66$ and $^{7/5}E^0 = 0.61$, see Figure 7.7a. The master number $^{7/5}E^0 = \lim_{Ca_0 \to 0} \frac{I_{7/5}}{Ca\lambda^{1.66}}$ as simulated by the MM-B-model for different droplet size distributions, $\Psi(R)$, and viscosity ratios, $\lambda > 3.5$, is displayed in Figure 7.7b. To infer $\langle R \rangle_{54}$, Equation (7.15) was used by exchanging E^0, p and the moment of radius as shown below:

$$\frac{^{7/5}Q_0}{\omega_1^2} = 0.61 \lambda^{1.66} \frac{\eta_m^2 \langle R \rangle_{54}^2}{\Gamma^2} . \tag{7.17}$$

In the experimental section 7.7.5 a correlation between the width of the distribution and the ratio of the different moment of radii $PDI_{53} = \frac{\langle R \rangle_{54}}{\langle R \rangle_{43}}$ is found for commercial samples, see Figure 7.12b. The evidence of such a correlation further confirms that the moments $\langle R \rangle_{43}$ and $\langle R \rangle_{54}$ are a good choice for the analysis of emulsion characteristics with the different higher harmonic ratios $I_{5/3}$ and $I_{7/5}$, respectively. The here proposed analysis yields information about droplet size distribution and interfacial tension. It should be pointed out that the LAOS characterization in its current form, is not limited to low polydispersity indices ($PDI < 2$) as it was assumed in the Palierne model [Friedrich 95, Jacobs 99]. From a comparison of Figure 7.6 and 7.7 the closeness of the plateau values E_0 and $^{7/5}E_0$ as well as the viscosity ratio dependent exponents p led to the assumption of a general hypothetical Equation, which might be applicable to all higher harmonic ratios used to investigate the different moments of the droplet size distribution. The plateau value and the exponent p were adapted from the $I_{5/3}$ simulation, which were less affected by the onset of numerical round off errors, thus are more precise at small Ca, see Figure 7.7c.

$$\frac{^{n+2/n}Q_0}{\omega_1^2} = {^{n+2/n}}E^0 \lambda^p \frac{\eta_m^2 \langle R \rangle_{\frac{n+1}{2}+2;\frac{n+1}{2}+1}^2}{\Gamma^2} \tag{7.18}$$

where $^{n+2/n}E^0 = 0.64$ and $p = 1.63$. In this article this general expression was applied for $n = 3, 5$ or $I_{5/3}$ and $I_{7/5}$, respectively. Extended simulations for superior higher harmonic ratios requires an improvement in the accuracy of the numerical simulations, to reduce the onset of numerical round-off errors occurring at small Ca and can be done in the future.

7.6.1 Introduction of a new polydispersity index PDI_{53}

The results from Section 7.5 and 7.6 have shown, that two distinct moments of the droplet size distribution can be determined with the nonlinear mechanical emulsioncurve E^0 and $^{7/5}E^0$, namely $\langle R \rangle_{43}$ and $\langle R \rangle_{54}$, respectively. The characterization of two moments of a distribution gains information about the width of the distribution, which will be expressed by a new defined polydispersity index PDI_{53}, derived as follows:

$$PDI_{53} = \sqrt{\frac{^{7/5}Q_0}{^{5/3}Q_0}} = \frac{\langle R \rangle_{54}}{\langle R \rangle_{43}} = \frac{\frac{\int R^5 \Psi(R)\,dR}{\int R^4 \Psi(R)\,dR}}{\frac{\int R^4 \Psi(R)\,dR}{\int R^3 \Psi(R)\,dR}} = \frac{M_5 M_3}{M_4^2} \tag{7.19}$$

where $\Psi(R)$ is the normalized droplet distribution ($\Psi(R) = \frac{n(R)}{\int n(R)\,dR} = \frac{n(R)}{M_0}$). Here, some analogies between this new scalar and the well known polydispersity index (PDI) to measure the colloidal distribution width are demonstrated [Stieß 09, Odian 04]:

$$PDI = \frac{\langle R \rangle_{43}}{\langle R \rangle_{10}} = \frac{\frac{\int R^4 \Psi(R)\,dR}{\int R^3 \Psi(R)\,dR}}{\int R \Psi(R)\,dR} = \frac{M_4}{M_1 M_3}, \tag{7.20}$$

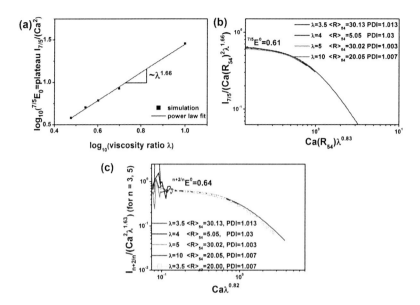

Figure 7.7: a) The simulation of $I_{7/5}$ using the conditions described in Figure 7.6 ($\omega_1/2\pi = 0.1\,\mathrm{Hz}$, $\eta_m = 1\,\mathrm{Pas}$, $R = 0.1 - 100\,\mu\mathrm{m}$, $\Gamma = 5.4\,\mathrm{mN/m}$ and $\lambda = 2.5 - 10$) and including different droplet size distributions resulted in viscosity ratio dependent plateau values, $^{7/5}E^0$. With the power law fit $^{7/5}E_\lambda^0 = ^{7/5}E^0\lambda^p$ with $p = 1.66$ and $^{7/5}E^0 = 0.61$, a superposition of the curves was achieved (b), once the averaged droplet radius $\langle R \rangle_{54}$ replaced the volume average radius $\langle R \rangle_{43}$ from the $I_{5/3}$ analysis in the capillary number. c) Superposition of $I_{5/3}$ (square symbol) and $I_{7/5}$ (solid lines) simulations show that the joint plateau value of $^{n+2/n}E^0 = 0.64$ was valid for both higher harmonics ratio. The simulation for $I_{5/3}$ is more precise at small Ca, due to less numerical noise. The discrepancy at large Ca between the curves of $I_{5/3}$ and $I_{7/5}$ had no impact on the analysis of the emulsion properties via the plateau value $^{n+2/n}E^0$ with $n = 3$ and 5 used within this thesis.

which is usually exploited as a measure of the distribution width. It is easy to demonstrate that both scalars defined in Equation (7.19) and (7.20) are greater than one and they tend to the unit value as the droplet population becomes monodisperse.

Some relationship between Equation (7.19) and (7.20) can be analytically found for most of the statistical distributions. As an example, we investigate the features of these indexes for two typical distributions, i.e. the lognormal and the normal (Gaussian) distribution.

7.6.1.1 Lognormal distribution

The single lognormal distribution is often encountered in the literature as a reasonable approximation for the droplet distribution. It depends on two parameters, the mean value $\mu \in \Re$ and the scale parameter $\sigma_d \geqslant 0$:

$$\Psi(R; \mu, \sigma_d) = \frac{1}{\sqrt{2\pi}\sigma_d R} \exp\left(\frac{-(\ln(R) - \mu)^2}{2\sigma_d^2}\right), \quad R \geqslant 0 . \tag{7.21}$$

It is worth mentioning, that $\Psi(R)$ degenerates to a monodisperse distribution when $\sigma_d \to 0$, and its variance increases with σ_d. An analytical expression for the moments of $\Psi(R)$ is available:

$$M_k = \int_0^\infty R^k \psi(R; \mu, \sigma_d) dR = \exp\left(k\mu + \frac{k^2\sigma_d^2}{2}\right) . \tag{7.22}$$

By exploiting Equation (7.22), PDI and PDI_{53} can be analytically derived:

$$PDI = \frac{M_4}{M_3 M_1} = \frac{\exp\left(4\mu + 16\sigma_d^2/2\right)}{\exp\left(\mu + \sigma_d^2/2\right)\exp\left(3\mu + 9\sigma_d^2/2\right)} = \exp\left(3\sigma_d^2\right) \tag{7.23}$$

$$PDI_{53} = \frac{M_5 M_3}{M_4^2} = \frac{\exp\left(5\mu + 25\sigma_d^2/2\right)\exp\left(3\mu + 9\sigma_d^2/2\right)}{\exp\left(4\mu + 16\sigma_d^2/2\right)^2} = \exp\left(\sigma_d^2\right) . \tag{7.24}$$

As a first remark one should notice that both indexes are increasing function of the only scale parameter σ_d. In addition, (i) the property $PDI \geq PDI_{53}$ is always satisfied and (ii) both indexes tend to unity as σ_d goes to zero.

7.6.1.2 Gauss distribution

In the case of the normal (Gaussian) distribution also for multimode distributions:

$$\Psi(R; \mu, \sigma) = \frac{1}{\sqrt{2\pi}\sigma_d} \exp\left(-\frac{(R-\mu)^2}{2\sigma_d^2}\right), \quad R \in \Re, R > 0 \tag{7.25}$$

there is still the dependence on two parameters: the mean value $\mu \in \Re$ and the standard deviation $\sigma_d \geqslant 0$. The scalars PDI and PDI_{53} can be analytically evaluated and expanded in terms of Taylor series with respect to the parameter σ_d:

$$PDI = -\frac{2\mu^2}{3\left(\mu^2 + 3\sigma_d^2\right)} + \frac{\sigma_d^2}{\mu^2} + \frac{5}{3} = 1 + \frac{3\sigma_d^2}{\mu^2} + O\left(\sigma_d^4\right) \tag{7.26}$$

$$PDI_{53} = \frac{\mu^2\left(\mu^2 + 3\sigma_d^2\right)\left(\mu^4 + 10\mu^2\sigma_d^2 + 15\sigma_d^4\right)}{\left(\mu^4 + 6\mu^2\sigma_d^2 + 3\sigma_d^4\right)^2} = 1 + \frac{\sigma_d^2}{\mu^2} + O\left(\sigma_d^4\right) . \tag{7.27}$$

Thus, PDI and PDI_{53} are straightforwardly related to the variance σ_d^2. It should be remarked that, once more: (i) always $PDI \geq PDI_{53}$ and (ii) $\lim_{\sigma_d \to 0} PDI = \lim_{\sigma_d \to 0} PDI_{53} = 1$. The suggested parameter PDI_{53} is, therefore, evaluated to be a quantity which is able to measure the width of the distribution.

7.7 Evaluation of the nonlinear mechanical master number

7.7.1 Materials and sample preparation

Dilute model emulsions

For the purposes of this thesis, model emulsions with and without diblock copolymer as compatibilizer were investigated. Two compounds were investigated where the dispersed phase is held constant. The first type of model blend consisted of a matrix of polyisobutylene (PIB) purchased from Biesterfeld Spezialchemie GmbH (Indopol H300, $M_n = 1.3\,\mathrm{kg/mol}$) and a dispersed phase of polydimethylsiloxane (PDMS) supplied by C.H. Erbslöh (Rhodorsil 47V 300000, $M_n = 150\,\mathrm{kg/mol}$). The commercially available diblock copolymer PIB-b-PDMS was purchased by Polymer Standards Services (PSS), Mainz, Germany and had a PIB block molecular weight of $3.4\,\mathrm{kg/mol}$ and PDMS block molecular weight of $14\,\mathrm{kg/mol}$ as provided by the manufacturer. The second type of model blend consisted of a matrix of polyisoprene. To vary the interfacial tension a diblock copolymer consisting of polyisoprene and polydimethylsiloxane was synthesized, see Chapter 8. The resulting diblock copolymer PI-b-PDMS consists of a PI block with $M_n = 3.1\,\mathrm{kg/mol}$ and a PDMS block with $M_n = 11.7\,\mathrm{kg/mol}$. The molecular weight of the matrix homopolymer PI was varied within the different polymer blends to generate different viscosity ratios, see Table 7.1. The amount of compatiblizer in the corresponding polymer blend is detailed below and is related to the name of the samples. The PDMS/PI-1 is prepared with a PI with $M_n = 18\,\mathrm{kg/mol}$ (KL15, Kuraray Europe GmbH) and in the other two blends, PDMS/PI-0.5 and PDMS/PI-0.7, the higher viscous PI was used with a molecular weight of $M_n = 28\,\mathrm{kg/mol}$ (LIR30, Kuraray Europe GmbH) as specified by the manufacturer [Kuraray 11]. The resulting interfacial tensions and other relevant physical properties of both types of dilute emulsions are listed in Table 7.1.

All raw materials PIB, PDMS and PI exhibited pure Newtonian behavior as the viscosity was constant over the investigated shear rate range for the temperatures used in these

Table 7.1: Main physical properties of the PDMS/PIB and PDMS/PI system. The interfacial tension for the PDMS/PIB emulsions without compatibilizer were determined by pendant drop measurements. The emulsion with PI as matrix were only investigated with compatibilizer. Therefore, no interfacial tension is available for the uncompatiblized blend.

Polymer	Formula	Molecular weight, M_n (kg/mol)	Density (g/cm^3) at 25 °C	Viscosity (Pas) at 25 °C	Interfacial tension (mN/m) at 25 °C
PDMS	$[-Si(CH_3)_2O-]_n$	150	0.896	280	
PIB	$[-CH_2C(CH_3)_2-]_n$	1.3	0.973	68	3.48
PI_{18}	$[-CH_2C(CH_3)CHCH_2]_n$	18	0.930	34	
PI_{28}	$[-CH_2C(CH_3)CHCH_2]_n$	28	0.930	130	
PIB-b- PDMS	$[-CH_2C(CH_3)_2-]_n-b-$ $[-Si(CH_3)_2O-]_m$	3.4-b-14			see Table 7.2
PI-b- PDMS	$[-CH_2C(CH_3)CHCH_2]_n-$ $b-[-Si(CH_3)_2O-]_m$	3.1-b-11.7			see Table 7.2

experiments (LAOS measurements in this study typically used $\gamma_0\omega_1 = 6\,s^{-1}$ and a steady shear rate of up to $\dot\gamma = 10\,s^{-1}$).

The viscosity ratio $\lambda = \eta_d/\eta_m$ between the dispersed phase and matrix were $\lambda = 4.2$ for PDMS/PIB, 2.2 for PDMS/PI$_{28}$ and 8.2 for PDMS/PI$_{18}$, which were large enough to prevent droplet breakup for moderate deformations with $Ca < 0.6$ [Grace 82]. Coalescence and droplet-droplet interaction was prevented by keeping the volume fraction low ($\Phi_{Vol} = 10\,\%$). Therefore, the emulsion was assumed to be stable meaning that its droplet size distribution did not vary during the experiment.

As the molecular weights for each block in the diblock copolymers were below or only slightly above the theoretical entanglement molecular weight, there were no entanglements expected within the phase of the diblock copolymers or between the blocks and the corresponding homopolymer phase. The entanglement molecular weights were taken from literature, [Dealy 06, Watanabe 11], and are 12 kg/mol for PDMS, 6.7 kg/mol for PIB and

5 kg/mol for PI. The amount of diblock copolymer is quoted as a weight fraction of the dispersed phase. Therefore, when the amount of diblock copolymer was given as 0.5 %, 0.7 % and 1 %, this meant that the overall blend composition has 0.05 %, 0.07 % and 0.1 % of diblock copolymer, respectively, and 9.95 %, 9.93 % and 9.9 % of PDMS homopolymer, respectively. The low amount of either PIB-*b*-PDMS or PI-*b*-PDMS did not influence the Newtonian flow behavior of the neat PDMS.

For the experiments, 50 mL samples were prepared by mixing the two phases in a 100 mL beaker using a variety of mixing techniques to vary the emulsion droplet size distribution. For example, for the blends without diblock copolymer compatibilizers (neat polymer blend), three different droplet size distributions were prepared by stirring either electrically at a speed of 100 rpm or 300 rpm or by stirring manually with a spatula until a homogeneous white blend was achieved after 300 s. In addition, 50 mL samples of compatibilized polymer blends were prepared by first manually blending with a spatula the PDMS homopolymer with either the PIB-*b*-PDMS diblock copolymer or with the PI-*b*-PDMS and then blending in the corresponding homopolymer PIB or PI, respectively, again until a homogeneous white blend was achieved after 300 s.

For each emulsion, the droplet size distributions was evaluated using light microscopy where Table 7.2 summarizes the results for the volume averaged droplet radius and the polydispersity index (*PDI*). A standard light microscope, Zeiss Axiophot, that had an objective LD Epiplan with a 20 fold magnification and an ocular of 10 fold magnification was used to evaluate the droplet size distribution. For each sample 500 to 600 droplets were manually measured to achieve a proper description and adequate statistics.

The interfacial tension between PDMS and PIB was measured using the pendant drop method from Dataphysics (OCA 15EC) and was found to be $\Gamma = 3.54\,\mathrm{mN/m}$ at 23 °C. With the temperature coefficient differential $d\Gamma/dt$ approximated by $-0.03\,\mathrm{mN/m/K}$, which is the upper limit for immiscible homopolymer blends [Koberstein 90], an interfacial tension of $\Gamma_{PD} = 3.48\,mN/m$ was estimated for 25 °C. This result is in good agreement with other estimations previously reported in the literature for this specific system [Kitade 97].

For variation of the interfacial tension, the diblock copolymers were added. The diblock copolymers PIB-*b*-PDMS and PI-*b*-PDMS are chemically linked polymers of the two blocks

PIB and PDMS or PI and PDMS,respectively, which are in both cases not miscible homopolymers. Thus, in a polymer blend of exactly these homopolymers, the addition of the PIB-b-PDMS or PI-b-PDMS results in a lowering of the interfacial tension. As mentioned in Section 5.1, the diblock copolymer settles preferably at the interface, where the polymer chains of each block reach into the corresponding bulk polymer phase, acting as a phase intermediator. The use of a diblock copolymer present at the interface of the droplets of the corresponding polymer blend, reduced the interfacial tension in its role as compatibilizer, but then made measurements of the interfacial tension via pendant drop impossible because $\Gamma < 3\,\mathrm{mN/m}$. As an alternative, the Palierne model for emulsions with a $PDI = \frac{\langle R \rangle_{43}}{\langle R \rangle_{10}} < 2$ [Palierne 90, Palierne 91, Graebling 93a] was used instead

$$\Gamma_{Pal} = \frac{\eta_m \langle R \rangle_{43}}{4\tau_{rel}} \frac{(19\lambda + 16)\left[2\lambda + 3 - 2\Phi_{\mathrm{Vol}}(\lambda - 1)\right]}{10\,(\lambda + 1) - 2\Phi_{\mathrm{Vol}}(5\lambda + 2)} \ . \tag{7.28}$$

In Equation (7.28) τ_{rel} is the relaxation time, η_m is the matrix viscosity, Γ_{Pal} is the interfacial tension (determined by the Palierne model) and $\lambda = \eta_d/\eta_m$ is the viscosity ratio of the two components. For the uncompatibilized blends where the distributions were broader, the Γ_{PD} values from the pendant drop analysis were used. The uncompatibilized blend PDMS/PIB-Spa with a PDI of 1.84 fails the Palierne analysis by gaining an interfacial tension of 2.57 mN/m from relaxation measurements. The restriction of a narrow droplet size distribution seems to be more stringent than reported in literature. However, when the distribution was narrower ($PDI < 1.8$) results from the Palierne model clearly showed that interfacial tension decreased with increasing diblock copolymers concentration. For these calculations, the characteristic relaxation time for each emulsion was first determined using a frequency sweep ($\omega_1/2\pi = 0.1\,\mathrm{Hz} - 10\,\mathrm{Hz}$) at a strain amplitude of $\gamma_0 = 0.3$ in absolute value, where the third relative higher harmonic $I_{3/1}$ is smaller than $0.5 \cdot 10^{-4}$ [Graebling 93a]. The interfacial tension was then calculated using Equation (7.28), see Table 7.2.

In contrast to the polymer blend consisting of PDMS in PIB and its corresponding diblock copolymer, the addition of PI-b-PDMS to the PDMS/PI blends has an influence on the linear rheology. The presence of a diblock copolymer at the interface can induce a second relaxation time, τ_β, also called the interfacial relaxation [van Hemelrijck 04, van Hemelrijck 05]. The reason can be found in the literature and is explained by interfacial tension gradients [Stone 90, Milner 96, Li 97, Velankar 01, Jeon 03].

Table 7.2: Main physical properties of the emulsions PDMS/PIB and PDMS/PI with and without diblock copolymer. $\langle R \rangle_{4,3,M}$ is the volume averaged radius determined via microscopy and τ_{rel} is the relaxation time from linear rheological measurements used to calculate the interfacial tension Γ_{Pal} via the Palierne model. For the uncompatibilized blends. Γ_{PD} is obtained by pendant drop measurements. For these the Palierne model failed, due to the broad droplet size distribution with $PDI > 1.8$. Whereas the compatibilized blends have an interfacial tension of $\Gamma < 3\,mN/m$, which is too small for pendant drop measurements. The interfacial tensions of the compatibilized PDMS/PI blends were calculated with the nonlinear mechanical emulsion number E^0 by measuring the intrinsic nonlinear ratio $5/3 Q_0$, see Equation (7.29).

nomenclature	stirring	$\langle R \rangle_{4,3,M}$ (μm)	PDI	wt-% PIB-b-PDMS in PDMS	τ (s)	Γ_{Pal} (mN/m)	Γ_{PD} (mN/m)	$5/3 Q_0$	Γ_E (mN/m)
PDMS/PIB-Spa	spatula	5.10	1.84	0	/	/	3.48	0.026	3.45
PDMS/PIB-100	100 rpm	9.30	2.48	0	/	/	3.48	0.063	4.27
PDMS/PIB-300	300 rpm	7.90	4.65	0	/	/	3.48	0.042	5.02
PDMS/PIB-0.5	spatula	3.30	1.18	0.5	0.52	2.24	/	0.028	2.15
PDMS/PIB-0.7	spatula	4.07	1.41	0.7	0.82	1.75	/	0.042	2.17
PDMS/PIB-1	spatula	2.25	1.17	1	0.609	1.30	/	0.045	1.16
PDMS/PI$_{28}$ − 0.5	spatula	5.04	1.57	0.5	/	/	/	0.042	3.00
PDMS/PI$_{28}$ − 0.7	spatula	2.65	1.19	0.7	/	/	/	0.018	2.42
PDMS/PI$_{18}$ − 1	spatula	6.22	1.90	1	/	/	/	0.047	2.76

It was found that the interfacial stress gradient causes Marangoni stresses, which induces a re-establishing of a uniform interfacial stress state and thus generates a second relaxation. Whereas diffusion processes are assumed to be one order of decade slower for a relaxation time smaller than 10 s. With an increasing coverage of the droplet, (increasing diblock copolymer concentration at same droplet size), the concentration gradient increases, which generates a faster relaxation. The structural relaxation is a fast process and appears at high frequencies, whereas the interfacial relaxation is a slow process and is measured at low frequencies. With an increasing coverage, the two processes cannot be distinguished in the linear measurements, because the interfacial relaxation approachs the time of a structural relaxation. It was shown, that a fitting of the linear elastic modulus G' with the Palierne model considering two relaxation processes, showed more reliable results than considering only the structural relaxation as applied within this thesis [van Hemelrijck 04].

The measurement of the frequency dependent elastic modulus and the corresponding relaxation spectra are shown in Figure 7.8a and b, respectively, for the three samples with the $PDI < 1.8$ PDMS/PIB-0.5, PDMS/PIB-0.7 and PDMS/PIB-1 and in Figure 7.8c and d for PDMS/PI$_{28}$-0.5, PDMS/PI$_{28}$-0.7 and PDMS/PI$_{18}$-1. The plotting of the first moment of the relaxation spectra, that means as $H(\tau) \cdot \tau$ versus τ, amplifies the contribution of slow processes, like the interfacial relaxation mechanism. For the compatibilized PDMS/PIB emulsions only the structural relaxation process is detected, thus a reliable relaxation time τ_α for the Palierne analysis with Equation 7.28 is measured. The compatilized PDMS/PI blends show both relaxation times as figured out in the relaxation spectra in Figure 7.8d. The relaxation times in the compatibilized PDMS/PI blends are not well distinct and actually for a PI-b-PDMS concentration of 1 % the relaxation times collapse, see Figure 7.8d. Thus a defined value for the structural relaxation time τ_α cannot be found. Therefore, the Palierne analysis is not useful for the interfacial tension evaluation with Equation (7.28) for the compatibilized PDMS/PI blends. Therefore, for these systems the nonlinear mechanical master number for emulsions E^0 is used to determine the interfacial tension. With the microscope the volume averaged radius $\langle R \rangle_{43}$ was determined, which is used as an input parameter for the evaluation with the emulsion number, Section 7.5, to determine the interfacial tension:

$$\Gamma = \sqrt{\frac{E^0}{5/3 Q_0}} \cdot \omega_1 \eta_m \langle R \rangle_{43} \lambda^{p/2}. \tag{7.29}$$

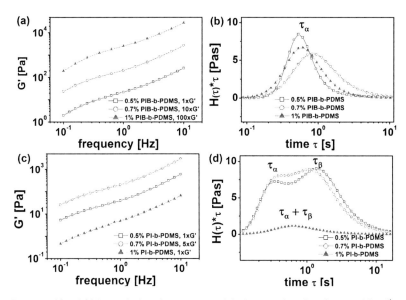

Figure 7.8: (a) and (c) Linear rheological measurement of the frequency dependent elastic modulus G' at 25 °C for compatibilized PDMS/PIB and PDMS/PI blends, respectively, in a cone plate geometry with 50 mm diameter. (b) and (d) Corresponding relaxation spectra for the compatibilized PDMS/PIB and PDMS/PI blends, respectively. The compatibilizer PI-b-PDMS generates an interfacial relaxation mechanism with relaxation time τ_β based on interfacial tension gradients on the droplet surface.

where $E^0 = 0.64$ and $p/2 = 0.82$. Together with the input parameters from the rheological experiment ω_1 as well as the measured matrix viscosity η_m and intrinsic nonlinear ratio $^{5/3}Q_0$, where the latter one is obtained by nonlinear FT-rheological measurements (Section 7.2), the interfacial tension can be calculated, see Table 7.2.

7.7.2 Coverage of the PDMS droplets in the matrix PI with compatibilizer PI-b-PDMS

The different relaxation mechanisms are a result of the partially covered droplet surface, which induces interfacial gradients and Marangoni stresses. In literature the coverage of PDMS droplets in a PI matrix is assumed to be 0.3 dbcp/nm², whereby dbcp is standing for diblock copolymer [van Hemelrijck 04]. This calculation assumes a lamellar structure with

Table 7.3: Coverage of the PDMS droplets with diblock copolymer PI-b-PDMS in the dilute PDMS/PI blends with Equation (7.30) where $M_{w,dbcp} = 15.4$ kg/mol. The value for the coverage c_0 of the droplets has the unit dbcp/nm^2 where dbcp is standing for *diblock copolymer*. The percentage value is given as a ratio of the total coverage with 0.3 dbcp/nm^2 to the coverage of the droplets [van Hemelrijck 04].

nomenclature	$\langle R \rangle_{43,M}$ (µm)	c_0 (bcp/nm^2)
PDMS/PI-0.5	5.04	0.0029 (0.98 %)
PDMS/PI-0.7	2.65	0.0022 (0.72 %)
PDMS/PI-1	6.22	0.0073 (2.42 %)

a lamellar thickness of 22 nm for a diblock copolymer of about 10 000 g/mol [Almdal 96]. Within this thesis the diblock copolymer PI-b-PDMS with $M_n = 14.8$ kg/mol is assumed to have a lamellar structure with a thickness of 23 nm detected by 2D SAXS measurements, see Section 8.5.4. The closeness of the values for the thickness with Almdal et al. allows the assumption of the same total coverage with 0.3 dbcp/nm^2. The coverage of the droplets within the different PDMS/PI blends can be calculated via [van Hemelrijck 04]:

$$c_0 = \frac{z \rho_d \langle R \rangle_{43} N_A}{300 M_{w,dbcp}} \tag{7.30}$$

where z is the fraction of the diblock copolymer relative to the amount of dispersed phase, ρ_d is the density of the dispersed phase, here 970 kg/L, N_A is the Avogadro number and $M_{w,dbcp}$ is the weight average molecular weight of the diblock copolymer. The $M_{w,dbcp}$ is determined by the $M_n = 14.8$ kg/mol of the NMR measurement and the $PDI_M = 1.04$ from the SEC-RI measurement to $M_{w,dbcp} = 15.4$ kg/mol, see Chapter 8. The percentage coverage with respect to total coverage can be found in Table 7.3. The smaller droplets in PDMS/PI-0.7 are less covered than the droplets in PDMS/PI-0.5 although the amount of compatibilizer is raised. With increased diblock copolymer in PDMS/PI-1 the coverage is remarkably increased, which also explains the coincide of the two relaxation mechanisms, due to a fastened interfacial relaxation process with τ_β.

Commercial emulsions

Two commercial water in oil emulsions, w/o-1 and w/o-2, were also investigated where the difference between the samples was that w/o-1 contained almond oil in addition to paraffin oil and w/o-2 did not. Each sample was first sheared once in a capillary rheometer Rheo Tester 2000 from Göttfert that had a round hole die of diameter 0.3 mm and length 30 mm. As the applied pressure changed, the shear rates ranged from $\dot{\gamma} = 6.7 \times 10^4 \, \text{s}^{-1}$ to $6.7 \times 10^5 \, \text{s}^{-1}$, which corresponded to piston speeds ranging from 1 mm/s to 10 mm/s using a teflon piston with 15 mm diameter. Table 7.4 summarizes the effect of shear on both the volume average droplet size and the polydispersity of the droplet size distribution compared to the original w/o emulsions. To clarify the notation, w/o-1-x means that the emulsion w/o-1 was sheared in the capillary rheometer at a piston speed of x (where the piston speed is taken instead of the shear rate due to the easier to specify values). As dynamic light scattering measurements (PSS Nicomp 380) require the sample to be diluted, the emulsions were diluted with isoparaffin to a concentration of 0.2 g/L. To test the effect of dilution, a series with decreasing dilution were measured to prove that the droplet size distribution was unaffected by the dilute concentration. To analyse the dynamic light scattering data, an inverse Laplace transformation was used when the distribution was very broad and/or bimodal as recommended by the commercial software. In all other cases a Gaussian analysis was used [Nicomp manual 11]. These results are included in Table 7.4.

For the nonlinear rheological experiments on the commercial samples, an excitation frequency of $\omega_1/2\pi = 1 \, \text{Hz}$ was chosen and the deformation amplitude varied in the range of $\gamma_0 \in [0.001; 0.3]$ in absolute values, see Figure 7.11.

7.7.3 FT-rheological measurements

The nonlinear mechanical emulsioncurve $^{n+2/n}E\left(Ca \cdot \lambda^{p/2}\right)$ and its limiting value $^{n+2/n}E^0$ for $n = 3$ and 5 as Ca approaches zero were introduced as a general description for emulsion properties such as $\langle R \rangle_{43}$ or $\langle R \rangle_{54}$, respectively, or the interfacial tension. The range of strain amplitudes was limited to the range where nonlinear behavior shows simple scaling law at small strain amplitudes within the constraints of the torque being measurable (50 nNm as specified by the manufacturer). The limit for the maximum strain amplitude is given by measurement artefacts, such as loss of sample due to centrifugal forces.

Table 7.4: Volume average diameter $\langle R \rangle_{43}$ and standard deviation σ_R of the commercial original and highly
sheared w/o-1 and w/o-2 emulsions. The results were determined using dynamic light scattering
with a PSS Nicomp 380 device. All the w/o-1 samples posess a bimodal distribution. For
bimodal distributions, the inverse Laplace transformation was applied, while for monomodal
narrow distributions, the Gaussian analysis was preferred. From the bimodal distributions, only
the average radius of the larger fraction (typically over 90 %) is listed with its standard deviation.

nomenclature	piston speed	shear rate $\dot{\gamma}$	$\langle R \rangle_{43}$ (nm)	σ_d (nm)
w/o-1	/	/	394	152
w/o-1-1	1 mm/s	$6.7 \times 10^4\,\mathrm{s}^{-1}$	384	48
w/o-1-5	5 mm/s	$3.3 \times 10^5\,\mathrm{s}^{-1}$	219	31
w/o-2	/	/	356	147
w/o-2-3	3 mm/s	$2.0 \times 10^5\,\mathrm{s}^{-1}$	283	110
w/o-2-7	7 mm/s	$4.7 \times 10^5\,\mathrm{s}^{-1}$	173	25
w/o-2-10	10 mm/s	$6.7 \times 10^5\,\mathrm{s}^{-1}$	183	50

The predicted strain amplitude dependencies of γ_0^2 and γ_0^4 for $I_{3/1}$ and $I_{5/1}$, respectively,
were measured for the different samples as shown in Figure 7.9. The specified regions
were characteristic for small amplitude behavior and were, therefore, analyzed immedi-
ately after the first significant increase above the noise level, N, which was caused by the
instrument resolution. Below the instrument torque resolution, I_n cannot be detected and
was therefore assumed to be equal to a constant value N. In the noisy region for small
strains and low torque, $I_{n/1}$ was a decreasing function that followed $I_{n/1} = \frac{N}{I_1} \propto \frac{N}{\gamma_0^1} \propto \gamma_0^{-1}$
if I_1 is still detectable. For sake of clarity, the noise and the onset of the γ_0^2 and γ_0^4,
respectively, dependence were fitted with a superposition of two physical phenomena. The
nonlinear behavior is described by a sum of the nonlinear behavior which is dominated
by the instrument noise at small strain amplitudes and by the scaling law behavior γ_0^{n-1}
(Equation (2.11)) at larger strain amplitudes. Thus, the strain amplitude dependent higher

harmonics can be described by the following superposition of the two scaling regions:

$$I_{n/1} = \underbrace{a_n \cdot \gamma_0^{-1}}_{\text{dominated by instrument noise}} + \underbrace{b_n \cdot \gamma_0^{n-1}}_{\text{dominated by nonlinear behavior}} \qquad (7.31)$$

where $n = 3$ or 5. Only the first increase in the measured higher harmonics was taken into account as this better corresponds to the scaling theory for small amplitude behavior. At higher strain amplitudes, asymptotic behavior to a plateau value was observed. It should be remarked that the onset of the power law behavior was seen at different γ_0 values for the two higher harmonics. The strain amplitude dependent $I_{n/1}$ curves, Figures 7.9 and 7.10 for the PDMS/PIB and PDMS/PI blends, respectively, the noisy region with $I_{n/1} \propto \gamma_0^{-1}$ with the determined fit parameter a_n and the region determined by the nonlinear behavior of the sample with $I_{n/1} \propto \gamma_0^{n-1}$ with the fit parameter b_n, have an intersection point where $a_n \gamma_0^{-1}$ is equal to $b_n \gamma_0^{n-1}$, i.e. at $\gamma_0 = \sqrt[n]{\frac{a_n}{b_n}}$. The intrinsic nonlinear $^nQ_0$ parameters are determined by the second term of the superposition, $I_{n/1} = b_n \cdot \gamma_0^{n-1}$ and results together with Equation (2.12) in:

$$^nQ_0 = \frac{I_{n/1}}{\gamma_0^{n-1}} = b_n \ . \qquad (7.32)$$

This means that the b_n values determined by the nonlinear fit of the strain amplitude dependent $I_{n/1}(\gamma_0)$ curves with $n = 3$ and 5, the intrinsic nonlinear ratio $^{5/3}Q_0$ is given by the ratio of exactly these fit parameters $\frac{b_5}{b_3}$. The determined intrinsic nonlinear ratios are used to determine the volume average values or the interfacial tensions as shown in Tables 7.5 and 7.2, respectively.

7.7.4 Dilute emulsions

Dilute emulsions consisting of either PDMS/PIB or PDMS/PI with or without the corresponding diblock copolymers PIB-*b*-PDMS and PI-*b*-PDMS, respectively, are characterized with the nonlinear mechanical master number E^0. Each sample has a different droplet size distribution and interfacial tension, see Table 7.2. PDMS/PIB emulsions prepared using a spatula instead of an electrical stirrer were more homogeneous as were the emulsions containing compatibilizer. LAOS measurements with an excitation frequency of $\omega_1/2\pi = 0.1\,\text{Hz}$ and a strain amplitude that varied from $\gamma_0 = 0.1$ to 10 in absolute values were performed to examine the higher harmonics $I_{n/1}$. A frequency sweep test performed at both the beginning and end of the LAOS measurement monitored the stability of the emulsion morphology. When the before and after dependencies for $G'(\omega)$ and

$G''(\omega)$ overlapped, within $\Delta G' < 3\%$, this was considered to be an indication of the emulsion microstructure stability [Carotenuto 08]. The strain amplitude dependent $I_{n/1}(\gamma_0)$ measurements are shown in Figures 7.9 and 7.10 for PDMS/PIB and PDMS/PI blends, respectively. Due to the inclusion of the noise region in the superposition of the two scaling regions γ_0^{-1} and γ_0^{n-1}, the measured data was also included below the minimum measured torque for $I_{3/1}$ and $I_{5/1}$. The polymer blend of PDMS in PI with diblock copolymer PI-b-PDMS has two relaxation times measured with the linear frequency sweep in Figure 7.8c and d. Therefore, the Palierne model could not be used to characterize the interfacial tension with Equation (7.28). Thus, in contrast to the PDMS/PIB blends, the intrinsic nonlinear ratio was used to determine the interfacial tension instead of the volume average radius, see Table 7.2.

For the PDMS/PIB blend the resulting values for the intrinsic nonlinear ratio, $^{5/3}Q_0$, evaluated using the nonlinear fit Equation (7.31) are collected in Table 7.5. Here interfacial tension was either measured by the pendant drop method or determined from linear frequency sweep measurements with the Palierne model. To calculate the droplet radius Equation (7.18) was transformed leading to the following explicit relationship for the volume average droplet radius:

$$\langle R \rangle_{43} = \sqrt{\frac{^{5/3}Q_0}{E^0} \frac{\Gamma}{\omega_1 \eta_m} \frac{1}{\lambda^{p/2}}} \; . \tag{7.33}$$

The terms on the right hand side are known from experiments, specifically the intrinsic nonlinear ratio $^{5/3}Q_0$ and Γ values (listed in Table 7.5), $\omega_1/2\pi = 0.1\,\mathrm{Hz}$, $\lambda = 4.1$, $\eta_m = 68\,\mathrm{Pas}$, $E^0 = 0.64$ and $p/2 = 0.82$. Using these values, the volume average radii was then estimated. Light microscopy analysis yielded a droplet size distribution that was averaged over 500 to 600 droplet radii to give a volume average radii designated as $\langle R \rangle_{43,M}$. Both $\langle R \rangle_{43,M}$ and $\langle R \rangle_{43,E}$ are shown in Table 7.5, and agreed well with each other. The shoulder in the strain amplitude dependent $I_{n/1}(\gamma_0)$ measurements are measured for samples with and without compatiblizer and, therefore, cannot be clearly related to interfacial phenomena like the Marangoni stresses, see Figures 7.9 and 7.10. Unfortunately, the derived correlation between $^{7/5}Q_0$ and the higher moment radius $\langle R \rangle_{54}$ cannot be tested for these dilute systems, because, while the resolution limit of the rheometer can detect the intrinsic behavior of $I_{5/1}$, it cannot resolve $I_{7/1}$ at small strain amplitudes. To illustrate this, the torque at the minimum of $I_{5/1}$ is shown in Figure 7.9a where this minimum torque was

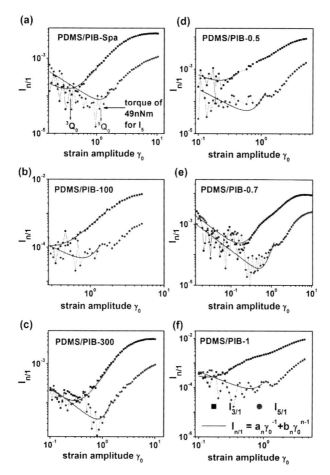

Figure 7.9: Measurements (symbols) of the higher harmonics $I_{n/1}(\gamma_0)$ with $n = 3, 5$ under LAOS for PDMS in PIB samples without compatibilizer (a-c) and with diblock copolymer (d-f). The use of the superposition of the two scaling regions γ_0^{-1} and γ_0^{n-1} with the nonlinear fit Equation (7.31) showed the transition from noise to significant intensity values. The nonlinear fit helped to determine the initial increase that followed a strain amplitude dependence of γ_0^2 and γ_0^4, respectively. The intersection point of the two curves with different slopes in Equation (7.31) defines the onset of the scaling law behavior at $\gamma_0 = \sqrt[n]{\frac{a_n}{b_n}}$ with $n = 3, 5$. The determined b_n values with $n = 3$ and 5 are the intrinsic nonlinear parameters $^3Q_0$ and $^5Q_0$. Thus, the ratio $\frac{b_5}{b_3}$ is the desired intrinsic nonlinear ratio $^{5/3}Q_0$. Additionally, an example for the minimum measured torque resolution for I_5 was added to (a), which confirmed the specifications by the manufacturer. The legend in (f) is valid for all diagrams.

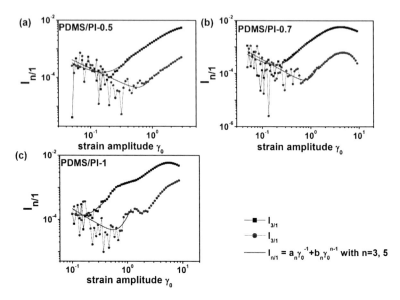

Figure 7.10: Strain amplitude dependent higher harmonic ratios $I_{n/1}$ for $n = 3$ and 5 of PDMS/PI with different amount of PI-b-PDMS diblock copolymer (symbols). All measurements were preformed in a cone plate geometry with 50 mm diameter and an angle of 0.4 rad. The temperature was set at 25 °C and controlled via a Peltier element. The nonlinear fit, Equation (7.31) yield the parameters a_n and b_n from which the nonlinear parameters $^3Q_0$ and $^5Q_0$ are determined for $^{5/3}Q_0$ (lines). The resulting intrinsic nonlinear ratio $^{5/3}Q_0$ was used to calculate the interfacial tension Γ_E, see Table 7.2, which could not be characterized by the Palierne model due to the interfacial relaxation time τ_α.

similar to reflects the absolute torque resolution (50 nNm) specified by the manufacturer.

7.7.4.1 Comparison between the polymer blends PDMS/PIB and PDMS/PI with compatibilizer

In the following the differences between the two polymer blends PDMS/PIB and PDMS/PI and the impact of the corresponding diblock copolymers PIB-b-PDMS and PI-b-PDMS, respectively, shall be summarized. The compatibilizer in PDMS/PIB has no influence on the structural relaxation mechanism with relaxation time τ_α. This is supported by the observation that within the measured frequency range, only a single relaxation time occurs,

Table 7.5: The radii values were determined via microscopy ($\langle R \rangle_{43,M}$) or via the nonlinear mechanical emulsionnumber E^0 ($\langle R \rangle_{43,E}$) for the model PDMS/PIB emulsions with and without diblock copolymer. In addition, $^{5/3}Q_0$ is shown from the nonlinear measurements, see Equation (7.33). The error of the volume averaged radius is estimated to be 10 % except for the PDMS/PIB-300 with 17 % difference possibly induced by the high polydispersity index $PDI > 4.5$, see Table 7.2.

nomenclature	$^{5/3}Q_0$	$\langle R \rangle_{43,M}$ (μm)	$\langle R \rangle_{43,E}$ (μm)
PDMS/PIB-Spa	0.026	5.10	5.14
PDMS/PIB-100	0.063	9.30	8.00
PDMS/PIB-300	0.042	7.90	6.50
PDMS/PIB-0.5	0.028	3.30	3.44
PDMS/PIB-0.7	0.042	4.10	3.29
PDMS/PIB-1	0.045	2.25	2.52

namely τ_α, even for low diblock copolymer concentrations, Figure 7.8. The interfacial relaxation with τ_β seems to be at least an order of magnitude slower than τ_α. The structural relaxation time appears unaffected by the interfacial elasticity and thus the Palierne model is a useful tool to calculate the interfacial tension as long as the polydispersity index is smaller than 1.8. For the emulsions with high polydispersity the newly invented characterization method based on FTR protocols is used to determine $\langle R \rangle_{43}$ or Γ. The here proposed analysis is based on large amplitude oscillatory shear and measures the strain amplitude dependent higher harmonics $I_{n/1}(\gamma_0)$ to determine the intrinsic nonlinear ratio. This ratio $^{5/3}Q_0$ was used to achieve the volume average radii of the PDMS/PIB blends which were not available by the Palierne analysis with its restriction to low polydispersities.

The interfacial relaxation for the PDMS/PI blend with PI-b-PDMS compatibilizer, is in the order of the structural relaxation time, τ_α, and therefore influences τ_α. The Palierne model, as used in Equation (7.28), is no longer appropriate, since the structural relaxation time is not clearly distinct from the interfacial relaxation. Therefore, the interfacial tension Γ of the model blend PDMS/PI with compatibilizer is characterized with the nonlinear mechanical master number for emulsions E^0. Other techniques like Wilhelmy plate or pendant drop measurements fail, due to high viscosities or low interfacial tensions, respectively.

The existence of both models, Palierne model and Maffettone Minale model in combination with the Batchelor theory, are necessary to fully characterize the dilute polymer blends, whereby the newly proposed nonlinear characterization has an extended application than the Palierne model in its actual applied form with respect to large polydispersity and appearance of an interfacial relaxation process.

7.7.5 Commercial emulsions

The two different commercial w/o-emulsions, w/o-1 and w/o-2, and their corresponding sheared samples, exhibited highly nonlinear behavior as it is shown in a frequency dependent spectrum in Figure 7.11a and in [Hyun 11] where w/o-2 was measured too. The measurement of $I_{n/1}$ where $n = 3, 5$ and 7 of the two different commercial w/o-emulsions, w/o-1 and w/o-2, and the corresponding sheared samples with the strain amplitude dependent $I_{n/1}(\gamma_0)$ is shown in Figure 7.11b-f . The nonlinear fit with Equation (7.31) was inserted to determine the onset of γ_0^{n-1} dependence. The intersection points b_n determined the $^nQ_0$ values for $n = 3, 5$ and 7. From these dependencies the nonlinear ratios $^{5/3}Q_0$ and $^{7/5}Q_0$ can then be determined. Since these are highly concentrated systems with $\Phi_{\mathrm{Vol}} \approx 0.75$, the interactions between the droplets cannot be neglected. Thus, the theory here developed cannot be fully applied, which means that an absolute value for $\langle R \rangle_{43}$ cannot be determined a priori. However a relative derivation of the droplet size using the intrinsic nonlinear ratio was developed instead. Figure 7.12a shows the linear dependence between $\langle R \rangle_{43}$ and $\sqrt{^{5/3}Q_0}$. A linear fit with $\sqrt{^{5/3}Q_0} = m \cdot \langle R \rangle_{43} + b$ using a slope of $m = -4.1$ and an intersection point of $b = 1639$ resulted in a correlation coefficient $\kappa = 0.96$, really close to the ideal value $\kappa = 1$ corresponding to perfect correlation. For radii above $700\,\mathrm{nm}$, the intrinsic nonlinear ratio $\sqrt{^{5/3}Q_0}$ was a constant value approximately equal to 100 in absolute values. Theoretical models considering the interaction between the dispersed phase particles are developed in the linear and steady state rheological regime for highly concentrated emulsions as they can also be used for foams with a high amount of gaseous dispersed phase, see also Chapters 1 and 9 [Princen 82, Princen 86]. Although the volume fraction is beyond the region, where isolated droplets can be consid-

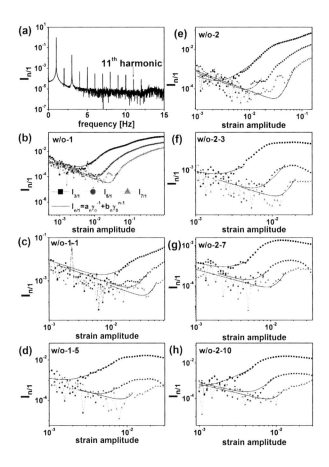

Figure 7.11: Commercial emulsions with a high volume fraction exhibited a highly nonlinear response under LAOS. Figure (a) shows the normalized intensity spectrum of w/o-2-10 measured at an excitation frequency of 1 Hz and a strain amplitude of 0.03. Already this small deformation gains 11 higher harmonics. An example for the maximum detected higher harmonic ever of concentrated emulsions is shown in Figure 3.5. In (b) the normalized higher harmonic intensities $I_{n/1}$ with $n = 3, 5$ and 7 (open symbols) show a strain amplitude dependence of γ_0^{n-1}, Equation (2.11), which can be fitted with the nonlinear fit of Equation (7.31). The superposition of the two scaling regions separates noise from the onset of the small strain amplitude behavior. The determined b_n values with $n = 3$ and 5 are the intrinsic nonlinear parameters $^3Q_0$ and $^5Q_0$. Thus, the ratio $\frac{b_5}{b_3}$ is the desired intrinsic nonlinear ratio $^{5/3}Q_0$. The legend in (b) is valid for all diagrams.

119

ered, it is assumed that the droplets are still spherical at rest, due to a volume fraction of $\Phi_{Vol} \approx 0.75$ and the polydispersity of the emulsions, see Table 7.4, which increase the maximum volume fraction for densely packed emulsions above $\Phi_{Vol} > 0.75$. An extension of existing theories into the nonlinear regime are not available so far, and are, therefore, not suitable to describe the large amplitude oscillatory shear behavior of the here investigated commercial emulsions. The empirical relation as described above is a first attempt to correlate the nonlinear rheological behavior to emulsion properties like volume average droplet size and droplet size distribution, see Figure 7.12.

As the behavior exhibited by these blends was highly nonlinear, $I_{7/1}$ could be clearly measured and analyzed, Figure 7.11. The nonlinear fit of Equation (7.31), was used with $n = 7$ to determine $^7Q_0$. As a result, information on the width of the distribution was found. The ratio of $\sqrt{\frac{7/5Q_0}{5/3Q_0}}$ was related to the standard deviation of the radius distribution, measured with dynamic light scattering, $\sigma_d = \sqrt{\langle R^2 \rangle - \langle R \rangle^2}$ with $\langle R^2 \rangle$ defined as the second moment and $\langle R \rangle^2$ as the first moment, respectively, [Jondral 02], of the distribution, Figure 7.12b. Overall, the ratio $\sqrt{\frac{7/5Q_0}{5/3Q_0}}$ was found to decrease with increasing width of the distribution. For the bimodal distributions, one representative value for the width of the distribution was necessary, therefore, the smaller droplet fraction was not included, as it was lower typically $< 10\,\%$. Although the total width of the distribution is thus automatically narrower compared with those of the w/o-2 samples, which were not bimodal, a linear fit of the double logarithmic scaling in Figure 7.12b results in a correlation coefficient of 0.74. The reason is that the low amount of small droplets does not influence significantly the nonlinear behavior, otherwise the w/o-1 measured data would not match to the w/o-2 data. Figure 7.12c and d are added to highlight the new invented correlation between the nonlinear parameter $\sqrt{5/3Q_0}$ and the volume average diameter as well as between the nonlinear ratio $\sqrt{\frac{7/5Q_0}{5/3Q_0}}$ and the standard deviation of the distribution, because an inverse dependence results in a worse correlation with coefficients only about 0.4.

The reason for the difference in the distributions for the w/o-1 and w/o-2 samples may be due to the use of almond oil in the w/o-1 samples. Almond oil consists of 95 % fatty acids, which are mostly oleic and linoleic acids [Burlando 10]. Compared to paraffin oil, which is a saturated hydrocarbon derivative, almond oil is more polar due to the ester group formed from the glycerine and the fatty acids, which could affect the droplet size distribution, even though the main ingredients are identical.

Comparison of Figures 7.12a and b shows that a piston speed of $> 2\,\mathrm{mm/s}$ corresponding to a shear rate greater than $1.3{\cdot}10^5\,\mathrm{s^{-1}}$ must be exceeded to measure a change in the volume average radius, but even the slowest applied shear had an effect on the width of the distribution. It was also observed that a piston speed of $7\,\mathrm{mm/s}$, corresponding to a typical shear rate of $\dot{\gamma} = 4.6{\cdot}10^5\,\mathrm{s^{-1}}$ created both the smallest average radius and the most narrow distribution. A higher shear rate did not continue this trend, but made the droplets coalesce and resulted in both a larger average radius and a larger standard deviation.

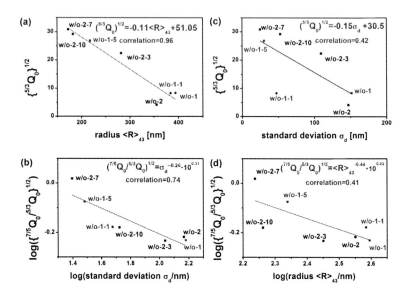

Figure 7.12: Commercial emulsions with a high volume fraction were analyzed to deduce the effect of the volume average radius on the intrinsic nonlinear behavior. It was found that, with increasing diameter, a decrease in $\sqrt{5/3 Q_0}$ was measured, (a). In (b) the relation between the width of the distribution, σ_d, and the ratio of $\sqrt{7/5 Q_0 / 5/3 Q_0}$ is displayed. The samples w/o-1 possess a bimodal distribution with a relatively low percentage of small droplets. For the average radius in (a) as well as for the standard deviation in (b), this smaller fraction of droplets was not considered, which is justified by the overall good correlation with coefficients close to 1. Whereas in (c) and (d) the dependencies of (a) and (b) were exchanged but results in a worse correlation with coefficients of only about 0.4. Thus, it can be summarized that $\sqrt{5/3 Q_0}$ is correlated to the volume average radius $\langle R \rangle_{43}$ and the ratio $\sqrt{7/5 Q_0 / 5/3 Q_0}$ is correlated to the width of the distribution.

8. Synthesis and characterization of the diblock copolymer PI-*b*-PDMS

In polymer research different polymerization techniques like step growth and chain growth are evolved and were first defined by Flory [Flory 53, Odian 04, Lechner 03, Tieke 05]. The major difference can be found in the polymerization degree-conversion ($P_n - U$) diagram, see Figure 8.1. Step growth means the stepwise reaction between functional groups of reactants. Generally the beginning of step growth polymerization builds oligomers, which react at the end to polymers. Thus, the P_n is in the beginning very low but increases at high conversion [Odian 04]. In contrast to this, the chain growth polymerization is the addition of monomers on reactive centers. Here, an initiation step is necessary. The chain growth reaches a high P_n in the beginning and runs into a plateau for high conversion. Within this thesis the living chain growth reaction is used to synthesize a diblock copolymer with a low polydispersity PDI_M defined as $PDI_M = \frac{M_w}{M_n}$. The $P_n - U$ diagram is visualized for comparison in Figure 8.1. Ionic synthesis have several advantages in contrast to other polymerization techniques. For example side reactions are less than in radical polymerization, where rearrangements are very common. Ionic synthesis also allows the synthesis of defined structures like a diblock copolymer with a narrow molecular weight distribution [Hsieh 96]. Further advantages can be found below. Within this thesis the anionic synthesis was used to generate PI-*b*-PDMS as compatibilizer for the PDMS/PI polymer blend. Therefore a short introduction into the theory of ionic synthesis with its main principles and characteristics will be made in the following

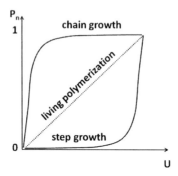

Figure 8.1: Diagram of polymerization degree P_n as a function of conversion U for step and chain growth. The living polymerization shows a different dependency explained in Section 8.1.

[Odian 04, Hsieh 96, Lechner 03, Tieke 05, Müller 09, Cowie 97].

8.1 Theory of ionic polymerization

Ionic synthesis is the umbrella term for anionic and cationic synthesis, where the polymerization is initiated via anions or cations, respectively. The present counter ion in the reaction plays an important role with respect to the polymerization rate and the resulting polymer structure [Hsieh 96]. The starting reaction can be formulated by the following two reaction schemes for anionic and cationic reactions, respectively, (8.1):

$$M + I^- \rightarrow IM_1^-$$
$$M + I^+ \rightarrow IM_1^+ \tag{8.1}$$

where M depicts the monomer and I the initiator. The initiation is either applied with nucleophiles or electrophiles for anionic or cationic synthesis, respectively, or by redox systems [Odian 04]. In the case of anionic initiation, the negative charge is stabilized by electron pulling groups, also called -I-effect or by resonance stabilization. For cationic synthesis, the positive charge is stabilized by the +I-effect of electron pushing groups or as well by resonance stabilization [Odian 04]. Monomers without these abilities are rather unsuitable for ionic synthesis. Therefore ethylene and propylene cannot be polymerized by ionic synthesis. The chain propagations can be described by the following route:

$$IM_1^- + nM \rightarrow IM_{n+1}^-$$

$$IM_1^+ + nM \rightarrow IM_{n+1}^+ \ . \tag{8.2}$$

As a characteristic of ionic polymerizations, the termination process can be regulated by the conditions prevailing in the polymerization process. Due to the electrostatic repulsion, termination by recombination, as it is found for example in radical polymerization, are not present. Chain end termination does occur in the presence of protic impurities or oxygen. The high demand of clean reaction vessels and of the chemicals used, increases the complexity of ionic polymerizations and thus the time and costs, which makes them less efficient from an industrial point of view. Szwarc defined this polymerization technique as a living polymerization, because the polymerization takes place until the monomer is depleted and the chains stay active for an indefinite time, [Szwarc 56, Szwarc 69]. Therefore the polymerization degree P_n is only determined by the initiator $[I]$ and monomer $[M]$ concentration, if the initiation process is quantitatively and much faster than the propagation [Hsieh 96]:

$$P_n = \frac{[M]}{[I]} \tag{8.3}$$

and is thus called a stochiometric polymerization [Lechner 03]. The advantages of a living polymerization are as follows:

- starting and propagation of the chains at the same time result in a narrow molecular weight distribution described by a Poisson distribution $P(x; \mu) = \frac{\exp(-\mu) \cdot (\mu^x)}{x!}$ where μ is the mean value [Consul 73]

- the polydispersity, PDI_M, is described by: $PDI_M = 1 + (P_n - 1)/P_n^2 \approx 1 + 1/P_n$ for a high polymerization degree P_n, which results in a PDI_M typically smaller than 1.1

- time and reaction agent of the termination is chosen by the experimenter and allows therefore the integration of functional end groups

- synthesis of diblock copolymers by addition of either a second monomer or a polymer, which is connected to the first living block

- the selective influence of the solvent and the temperature on the polymerization rate and polymer structure, as shown in Equation (8.4) and in Section 8.2, respectively.

The polarity of the solvent effects the equilibrium of ion pairs and determines the reaction rate, Equation (8.4).

$$
\overbrace{\underbrace{\left[P^-,E^+\right]_n}_{\text{aggregated ion pairs}} \rightleftharpoons \underbrace{nP^-,E^+}_{\text{contact ion pairs}}}^{\text{in nonpolar solvent}} \rightleftharpoons \overbrace{\underbrace{P^-/S/E^+}_{\text{solvent-separated ion pairs}} \rightleftharpoons \underbrace{P^- + E^+}_{\text{free ions}}}^{\text{in polar solvent}}
$$

$$
+M \downarrow k_{agg} \qquad + M \downarrow k_{\pm} \qquad + M \downarrow k_{\pm,S} \qquad + M \downarrow k_{-or+} \tag{8.4}
$$

where P^- and E^+ are standing either for the living chain or the counter ion with respect to anionic or cationic polymerization. S replaces the solvent and M the added monomer. Fuoss, [Fuoss 54], and Winstein [Winstein 54], proposed the existence of different ionic structures and their relation to the reaction rate. A loosely associated ion pair, like when they are separated by the solvent, is much more reactive than a tightly contacted ion pair. This is of course also valid for the free ions with a high reactivity in contrast to the opposed aggregated ions with nearly no reactivity. This is also reflected in the reaction rate constants where $k_{agg} \ll k_{\pm} \ll k_{\pm,S} < k_{-or+}$. In the case of a low solvation force the monomer is inserted into a covalent bond, which is also called a pseudoionic polymerization and insertion reaction, respectively, [Lechner 03].

The equilibrium between the aggregated ion pair and the free ions create an equilibrium between sleeping and active polymer chains. Only if the exchange between both species is fast compared to the reaction time, an equal propagation rate of all chains is secured, which at the end generates a homogeneous and nearly monodisperse distribution. Analog, the initiation has to be quantitatively as well as the distribution of the monomer has to be equally in the reaction system, that the propagation time and probability of all initiated monomers is the same.

Cationic polymerization is preferred by monomers with an electron pushing substituent, so that the electrophilic polymer chain with its positive charge reacts in a propagation step. In contrast, anionic polymerizations are supported by monomers with electron pulling substituents, which facilitate the nucleophilic addition of the negative charged polymer chain. Both reactions only take place at temperatures below the ceiling temperature T_c [Lechner 03, Dingenouts 10]. The diblock copolymer PIB-b-PDMS is commercially available and was purchased from the company Polymer Standards Service GmbH (PSS) Mainz, Germany. The diblock copolymer polyisoprene-block-polydimethylsiloxane (PI-b-PDMS) is synthesized within this thesis via anionic polymerization, which is described in details in the following.

8.2 Regioselective polyisoprene polymerization

Isoprene is a 1,3-diene with two stereoisomers in equilibrium, the cis and trans constitution in Figure 8.2, [1]. Out of these isomers, four different constitution structures are polymerized. A regioselective reaction is controlled via the polarity of the solvent. In nonpolar solvents 1,4-isomers are preferably generated, whereas polar solvents routes into 3,4 and 1,2-isomers, respectively, see Figure 8.2. The reason is the solvent separated ion pair, where the counter ion can delocalize. Thus the negative charge is delocalized from the α-C-atom to the γ-C-atom, while the γ-C-atom is preferred over the α-C-atom, due to the shielding of the α-position by the solvated cation. The nucleophilic attach is performed by the less sterically hindered γ-C-atom resulting in predominantly (1,2) and (3,4) configuration, Figure 8.3 [Lechner 03, Hsieh 96]. In contrast, nonpolar solvents with contact ion pairs result in mainly (1,4) polymerization. The regioselectivity of cisoid and transoid isoprene is controlled by an intermediate state with a Diels-Alder type six-membered ring activation complex between the cisoid-isoprene and the carbon-lithium bond of the active chain end, Figure 8.4. Therefore the reaction rate constant, k_p(cis), is about eight times faster than k_p(trans) [Worsfold 78].

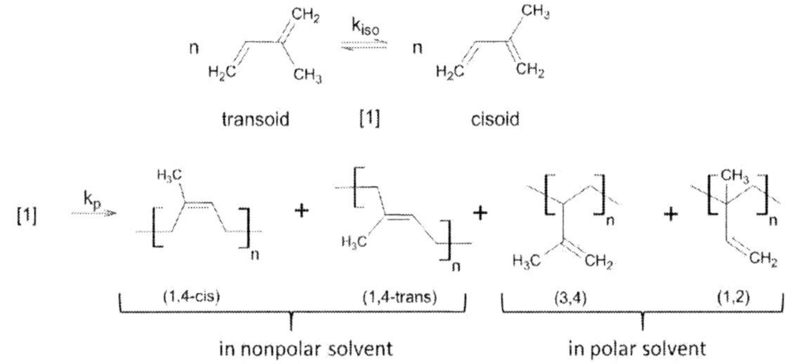

Figure 8.2: The upper reaction is the equilibrium of the monomer isoprene in its two stereoisomers transoid and cisoid [Lechner 03]. The polarity of the solvent controls the regioselectivity of the polymerization reaction.

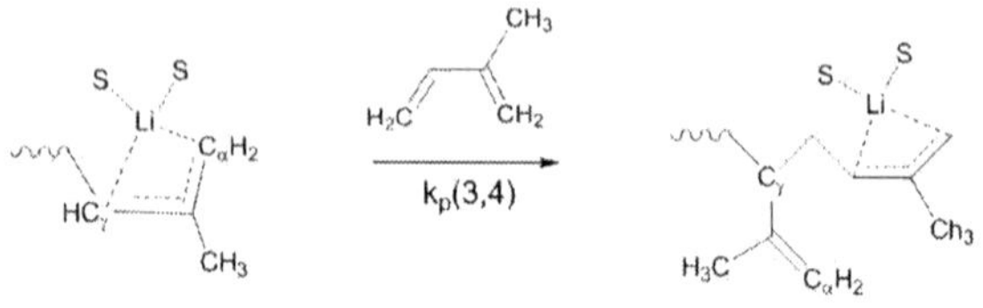

Figure 8.3: In polar solvents the (3,4) or (1,2) reaction, respectively, becomes dominant, due to the delocalization of the charge. The delocalization is induced by the solvent separated ion pairs with Lithium as counter ion and the shielding of the α-C-atom by the solvated counterion [Lechner 03] whereby S is figuring the solvent.

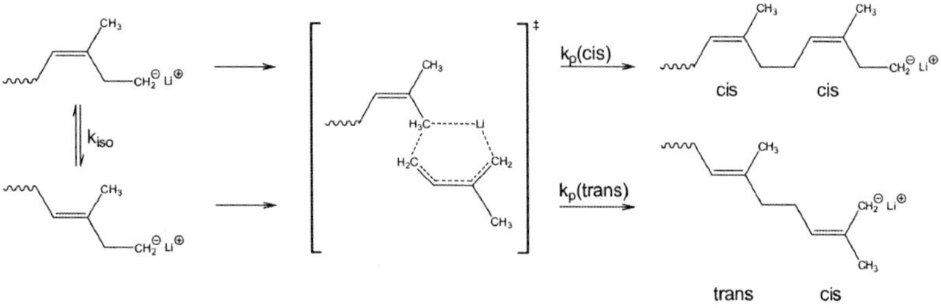

Figure 8.4: Reaction of a cis- and trans-polyisoprene with a cisoid monomer. The intermediate state with a Diels-Alder type six-membered ring activation complex is only possible with a cisoid-isoprene [Worsfold 78]. Therefore the reaction rate constant k_p(cis) is about eight times faster than k_p(trans).

8.3 Living ring opening polymerization

The polymerization of cyclic siloxanes is generally initiated by strong inorganic, organic or organometallic bases. This initiation step leads to a silanolate anion, which is the active center for the propagation step [Müller 09]. Generally the activated monomer exists in nonpolar solvents only as an ion pair, which is in equilibrium with higher aggregates. The aggregates are less reactive than the free ions, but are more selectively in the reaction. After the initiation period the polarity is increased by the addition of polar solvent to separate the ion pairs and with it the aggregates to induce the propagation step. The donor character of the solvent should not be too high, to minimize side reactions like backbiting, [Bellas 00]. The ring opening polymerization is a selective propagation step beside backbiting. Backbiting is the attack of the PDMS chains by their living ends and subsequent formation of deactivated unstrained rings. The cyclic monomer of PDMS possess a ring tension. Therefore, the ring opening reaction in the propagation step is enhanced as long as the activity of the living centers is not too high. If the reaction centers are too reactive their selectivity with respect to the desired propagation step with the monomer against backbiting with the existing PDMS chain is lost [Maschke 92]. In practice ring opening polymerizations are typically terminated at 75 % conversion, to avoid side reactions [Elkins 04]. Additionally termination agents are chosen differently from generating hydroxyl end groups, which could easily provide side reactions like backbiting and dehydration [Yilgor 98].

8.4 Synthesis of PI-*b*-PDMS

The diblock copolymer polyisoprene-b-polydimethylsiloxane (PI-*b*-PDMS) is synthesized in a sequential anionic polymerization utilizing high vacuum techniques to avoid impurification, of for example water and oxygen [Hadjichristidis 00]. The living polymerization allows a sequential polymerization where PI is synthesized until complete conversion of the monomer and afterwards the active chain end is propagated with the second monomer hexamethylcyclotrisiloxane (HMCTS) for the synthesis of PDMS [Bellas 00, Elkins 04]. The isoprene polymerization is started with sec-butyllithium (sec-BuLi) in toluene and the polymerization was allowed to proceed at room temperature. Due to the absence of any termination agents, the living chain PI can propagate with the added monomer HMCTS. The initiation of HMCTS proceeds for 12 h to assure a completed ring opening before the polar solvent THF is added to achieve solvent separated ions with a higher propagation rate. After a conversion of about 75 % the reaction was terminated with trimethylchlorosilane, Figure 8.5.

8.5 Characterization of PI-*b*-PDMS

The diblock copolymer is characterized with a variety of analytical methods like size exclusion chromatography (SEC), ^{1}H NMR and differential scanning calorimetry (DSC). While single methods do not suffice for diblock copolymer analysis, a combination of severall methods leads to a satisfactory characterization. In the following the analyzing techniques and several problems for this diblock copolymer PI-*b*-PDMS will be mentioned. Size exclusion chromatography in THF with a differential refractive index (RI) detector

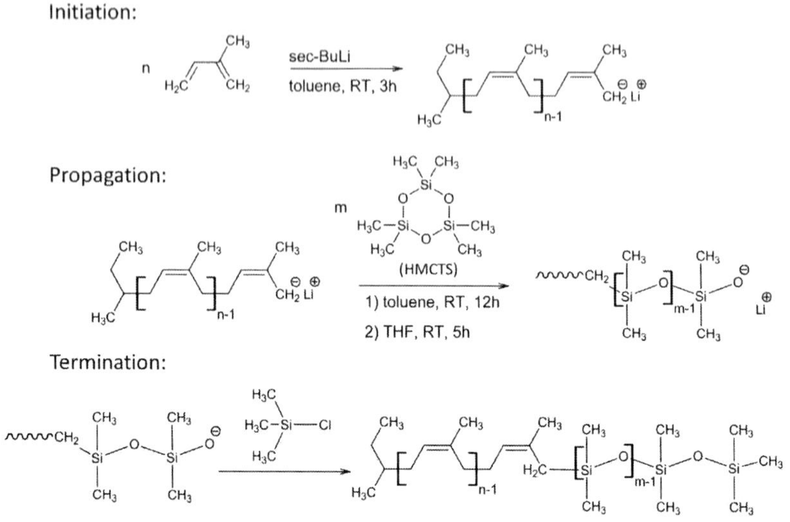

Initiation:

Propagation:

Termination:

Figure 8.5: Reaction scheme of the sequential anionic synthesis of polyisoprene and polydimethylsiloxane.

yields no information about the PDMS content due to the non pronounced difference of the refraction indices between THF and PDMS. Other detectors like the ultraviolet (UV) detector is neither available for PI nor for PDMS, due to nonexisting functional groups, which absorb in the UV wavelength range. The light scattering detector fails because the molecular weight of the diblock copolymer was too small. For the viscometer detector different molecular weights with the same ratio of the blocks is needed to determine the Mark Houwink parameters [Wu 05, Mori 99]. Using small angle X-ray scattering the whole diblock copolymer could be analyzed but no information about the single blocks is available. Using ^{1}H NMR characterization allows the calculation of the PDMS content but does not say anything about the complete conversion of the PI precursor to the diblock copolymer. The characterization of a diblock copolymer needs a well considered combination of characterization methods. Within this thesis several analyzing techniques were used. SEC with either a RI detector or with an infrared (IR) detector was used to determine the molecular weights of the precursor and the diblock copolymer. ^{1}H NMR spectroscopy was applied to determine the molecular weight of the second block PDMS. Differential scanning calorimetry (DSC) was used to determine the presence of a heterogeneity depending on a phase separation of the diblock copolymer. Additionally small angle X-ray scattering was used to determine the spatial distribution of the phase separated diblock copolymer.

A SEC chromatogram with RI detection is shown in Figure 8.6 together with the measurement of the PI precursor from the sequential polymerization. The weight averaged

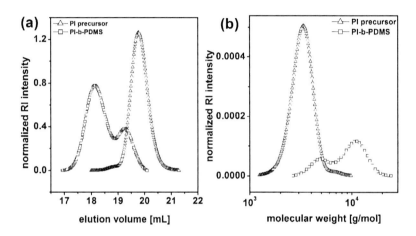

Figure 8.6: a) SEC of the PI precursor and the diblock copolymer in THF. The RI detector does not resolve a signal for PDMS, due to the low difference with the refractive index of THF. b) The weight averaged molecular weights are determined against a PI calibration. The M_w are 3.3 kg/mol for PI, 11 kg/mol and 4.8 kg/mol, respectively, for the bimodal signal measured with PI-b-PDMS. The polydispersity index ($PDI_M = \frac{M_w}{M_n}$) was found to be 1.07 for PI, and 1.04 and 1.07, respectively, for the two peaks with increasing elution volume of the diblock copolymer.

molecular weights determined from this elugram are 3.3 kg/mol for PI, 11 kg/mol and 4.8 kg/mol, respectively, for the bimodal signal measured with PI-b-PDMS. The bimodal distribution of the diblock copolymer cannot be quantitatively characterized, since it is not clear if the bimodal distribution corresponds either to a recombination termination reaction of the diblock copolymer or to homopolymer PI, which was terminated during the sequential polymerization. The detection of the PDMS block could clarify the uncertainty. A new measurement technique, which is based on the PhD thesis of Timo Beskers,

a group member of Professor Wilhelm, was applied for the diblock copolymer characterization. This project deals with the combination of SEC and Fourier Transform infrared
spectroscopy (FT-IR). The new detection method allows the simultaneous measurement
of PDMS and PI in the diblock copolymer.

8.5.1 SEC combined with FT-IR

The newly invented measurement technique of Beskers et al. [Beskers 10] combines SEC
with FT-IR detection, where a RI signal deals as reference signal for verification. This
method yields a measurement of a signal for PI as well as for PDMS and is therefore
advantageous for the characterization of the synthesized diblock copolymer. FT-IR spectroscopy is based on incident polychromatic light in the infrared region onto the sample.
The chemical environment of the molecular structure determines the absorption of characteristic wavelengths by the material if a fully optimized IR setup, a fully optimized
SEC column and a fully optimized sample cell are used in combination with a special,
mathematical solvent suppression as it can be found in Beskers et al. [Beskers 11]. The
combined measurement of SEC and IR detection (SEC-IR) yields a 2D chromatrogram
with respect to the elution volume and the IR vibrational bands of the polymeric sample,
see Figure 8.7a. Out of this chromatogram the characteristic vibrational bands for PDMS
and PI are determined and plotted as function of the elution volume, Figure 8.7b and c
for PDMS and PI, respectively. PDMS exhibit absorption at 1010-1050, 1100-1126 and
1265-1275 cm^{-1}. The PI vibrational bands are hidden beyond the water signal. With
subtraction of the water signal, the characteristic IR-bands of PI are extracted at the
wavenumbers 1371-1403 cm^{-1} and 1426-1455 cm^{-1}. Figure 8.7c shows additionally the RI
signal of PI for comparison.

The PDMS signal shows a monomodal distribution, whereas the PI analysis shows a shoulder in the IR and RI detection. The shoulder at higher elution volume is thus clearly
related to PI homopolymer, which could be generated by a termination reaction due to
impurities affected by the addition of HMCTS in the sequential polymerization, Section
8.4. This means the bimodal distribution measured for the diblock copolymer in Figure
8.6 is not related to a recombination product of the diblock copolymer, but appears, due
to PI homopolymer. From the PDMS chromatogram the molecular weight of the diblock
copolymer is determined against a PS calibration. The value for M_w of 10.2 kg/mol is close

to the determined weight average molecular weight of 11 kg/mol with the PI calibration. A deeper focus on the accuracy and thus applicability of a PS calibration for polyisoprene as well as for the diblock copolymer is not further explored and discussed here. The reason is, that the molecular weights determined with the two calibration curves, PS and PI, are within the typical 10 % error of SEC measurements. During the sequential anionic polymerization some PI polymer chains were terminated. The deactivated chains cannot react with the added second monomer, thus have a lower molecular weight than the resulting diblock copolymer after complete synthesis and are measured at a higher elution volume in the SEC elugram. The sequential anionic synthesis allows the SEC measurement of the first synthesized block, here PI. This measurement is added in Figure 8.8 to explore the appearance of the shoulder in the IR and RI signal at the characteristic wavelength of PI. The shoulder is measured around 55.5 mL, which is related to a molecular weight of $M_p = 5.6$ kg/mol. The precursor homopolymer PI, measured with the same equipment, has an elution volume of 58.7 mL which is related to a molecular weight of $M_p = 2.8$ kg/mol, both based on the same PS calibration. Two times the M_p of 2.8 kg/mol of the precursor equals 5.6 kg/mol measured with the diblock copolymer. This means, that the shoulder in the diblock copolymer signal is assumed to be due to a recombination termination reaction of the first synthesized block PI. The amount of homopolymer PI in the diblock copolymer can be calculated from the ratio of the integrals of each peak of the IR signals for PI and PDMS, see Figure 8.8b. Each peak is fitted with a Gaussian curve, see Equation (7.25). The PDMS signal is chosen, because the integral is not affected by the PI homopolymer but determined by the diblock copolymer. The integrals of the peaks for PI and PDMS are adjusted to the number ratio of the NMR intensities $\frac{PI}{PDMS} = \frac{0.23}{0.77}$, see Equation (8.8). The PI homopolymer is determined by the ratio of the two integrals of the PI signal to $\frac{0.075}{0.165} = 0.45$. This means 45 number% of PI homopolymer with respect to the PI in the diblock copolymer. Whereas with respect to the whole diblock copolymer the amount of PI homopolymer yields only $\frac{0.075}{0.75+0.165} = 0.08$, thus 8 %.

The extensive characterization is achieved by the combination of all SEC measurements, SEC-IR and SEC with RI detection of the diblock copolymer as well as of the PI precursor. The molecular weight of the PI block determined with the PI calibration is used for the determination of the block molecular weight of PDMS in Section 8.5.2. The molecular weight of the PI precursor and the PI homopolymer in the diblock copolymer show in

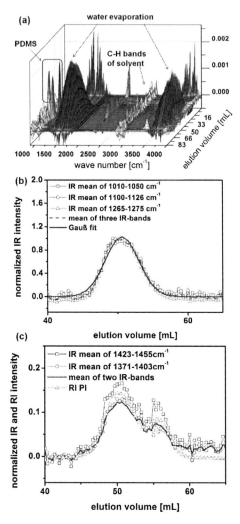

Figure 8.7: a) Size exclusion chromatogram with infrared spectroscopy detector. The IR-bands for the PI are covered by water peaks. The PDMS shows three characteristic IR-bands between 1010-1050, 1100-1126 and 1265-1275 cm^{-1}. The PI characteristic bands can be found at 1371-1403 and 1426-1455 cm^{-1} if the water signal is mathematically suppressed [Beskers 10]. b) Elugram of the PDMS at the three characteristic IR-bands. The solid line represents the mean signal of the three signals and has a maximum at 50 mL. The RI signal is not available for PDMS. The fit with a Gauss (compare Equation (7.25)) yields a standard deviation $\sigma_d = 4.7$ mL and a mean value $\mu = 50.6$ mL c) Elugram of the PI at two characteristic IR bands in comparison with the RI signal.

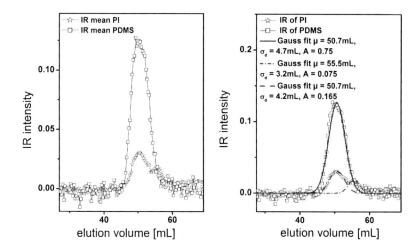

Figure 8.8: (a) Size exclusion chromatogram with peak areas adjusted to the relative NMR intensity ratio $\frac{PI}{PDMS} = \frac{0.23}{0.77}$ for the IR signals of PI and PDMS. (b) The shoulder in the IR signal of PI is fitted with a Gauss function to determine the related integral A. In the following the ratio of the areas of the Gauss fits for the two peaks of the IR signal of PI and the IR signal of PDMS are used to determine the amount of PI homopolymer. The PI signal consists of 45 number% PI homopolymer with respect to PI in the diblock copolymer. In comparison to the whole diblock copolymer the PI homopolymer yields 8 number%. The precursor PI (in (a)) has a half molecular weight ($M_p = 2.8\,\mathrm{kg/mol}$) than the hompolymer PI in the diblock copolymer (shoulder in (b) $M_p = 5.6\,\mathrm{kg/mol}$), both based on a PS calibration.

comparison, that recombination of PI during the sequential anionic synthesis is assumed to be most probably, because of two reasons. The first one is that the PI molecular weight was doubled from 2.8 to 5.6 kg/mol and the second one is that PDMS could only be measured for molecules of higher masses (smaller elution volume). Accordingly, the bimodal distribution in Figure 8.6 appears due to the presence of PI homopolymer and not due to a recombination product of the diblock copolymer.

The results from the SEC-IR measurement in comparison with SEC measurements and RI detection will be shortly summarized and are combined in Table 8.1 together with the following NMR analysis. It was found, that the diblock copolymer has a molecular weight of $M_w = 11\,\text{kg/mol}$ with respect to a PI calibration. The second peak at higher elution volume could be related to PI homopolymer. Additionally it was assumed, that the PI homopolymer is generated via a recombination termination reaction. The amount of PI homopolymer was determined to be 8 number% of the synthesized diblock copolymer. The block of PI can be determined from the SEC measurement of the precursor and has an M_w of $3.3\,\text{kg/mol}$ with respect to a PI calibration. As a last point, it should be mentioned, that the IR detector could measure a signal for the PDMS block, which was not available with other detectors like UV-, light scattering- or viscometer detector for different reasons mentioned in the beginning of this Section.

8.5.2 ^{1}H NMR characterization

The SEC analysis in combination with a ^{1}H NMR measurement is a useful tool to explore the proportionality and the exact molecular weights of the blocks. The SEC is sensitive to the sizes of the blocks, whereas the NMR measurement gains information about the chemical environment of the protons, whereby the intensities exhibit additional information about the ratio of protons at each C-atom. For further analysis, the integrals of the peaks, characteristic for each block, were normalized to one proton. For (1,4)-PI the single proton at C_3, Figure 8.9, of cis and trans conformation is detected at 5.10 ppm, whereas the intensities of C_1 with two protons of (3,4)-PI at 4.64-4.74 ppm are negligible. PDMS has a peak at 0.05 ppm, which is generated by six protons of each siloxan unit. Therefore the integral for a single proton are given for each block by:

$$I(\text{PI}) = \frac{5.26}{1} = 5.26$$
$$I(\text{PDMS}) = \frac{57.16}{6} = 9.52 \; . \tag{8.5}$$

Considering that the intensity of the PI also contains the contribution of the 45 number% PI homopolymer. The intensity for $I(\text{PI})$ changes to:

$$I(\text{PI}) = \frac{5.26 \cdot 0.55}{1} = 2.89 \; . \tag{8.6}$$

The normalized overall intensity is

$$I_{\text{ges}} = 2.89 + 9.52 = 12.41 \tag{8.7}$$

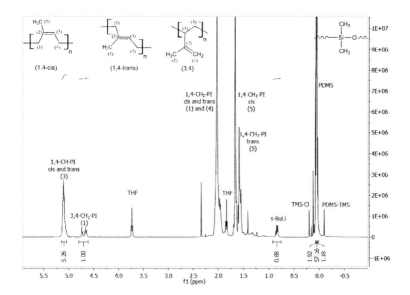

Figure 8.9: ^{1}H NMR spectrum of an anionically synthesized PI-b-PDMS with molecular weights of 3.3 kg/mol of PI and 6.4 kg/mol of PDMS (400 MHz, $NS = 1024$, CDCl$_3$, T=25 °C).

and the relative ratios of PI and PDMS are given by

$$\text{PI: } \frac{2.89}{12.41} = 0.23$$
$$\text{PDMS: } \frac{9.52}{12.41} = 0.77 \ . \tag{8.8}$$

With the molar masses of isoprene (68 g/mol) and the repetition unit of PDMS (75 g/mol) the mass ratio of PI and PDMS are calculated:

$$\text{mass ratio of PI} = 0.23 \cdot 68 \, \text{g/mol} = 15.64 \, \text{g/mol}$$

$$\text{mass ratio of PDMS} = 0.77 \cdot 75 \, \text{g/mol} = 57.75 \, \text{g/mol}$$

$$\text{total mass ratio} = 73.39 \, \text{g/mol}$$

$$\tag{8.9}$$

which results in the following percental mass ratios:

$$\text{PI: } \frac{15.64}{73.39} = 0.21$$
$$\text{PDMS: } \frac{57.75}{73.39} = 0.79 \tag{8.10}$$

With the densities of PI ($\rho = 0.909\,\mathrm{g/cm^3}$) [Brandrup 99] and of PDMS ($\rho = 0.896\,\mathrm{g/cm^3}$) [Brandrup 99] the volume fraction of each block is calculated to:

$$\text{volume ratio of PI} = \frac{15.64\,\mathrm{g/mol}}{0.909\,\mathrm{g/cm^3}} = 17.21\,\mathrm{cm^3/mol}$$

$$\text{volume ratio of PDMS} = \frac{57.75\,\mathrm{g/mol}}{0.896\,\mathrm{g/cm^3}} = 64.45\,\mathrm{cm^3/mol}$$

$$\text{total volume ratio} = 81.66\,\mathrm{cm^3/mol}$$

$$(8.11)$$

which results in the following percental volume ratios:

$$f_{\mathrm{PI}} : \frac{17.21}{81.66} = 0.21$$

$$f_{\mathrm{PDMS}} : \frac{64.45}{81.66} = 0.79 \qquad (8.12)$$

From the SEC measurements of the precursor PI the number average molecular weight is calculated from the weight average $M_w = 3.3\,\mathrm{kg/mol}$ and the PDI_M of 1.07. M_n is determined to be 3.1 kg/mol. Thus, for the PDMS block follows:

$$M_n(\mathrm{PDMS}) = \frac{0.79}{0.21} \cdot 3.1\,\mathrm{kg/mol} = 11.7\,\mathrm{kg/mol}\,. \qquad (8.13)$$

The total mass is confirmed by the standard SEC measurement with RI detection in THF of the diblock copolymer, Figure 8.6. The weight averaged molecular weight was measured to be 11.0 kg/mol with a PI calibration, because the second peak at higher elution volume was determined to belong to PI hompolymer. The PDI_M for the diblock copolymer is 1.04 and thus the number averaged molecular weight 10.6 kg/mol. The number averaged molecular weight M_n, calculated from the single blocks contribution from the NMR measurement, was determined to be 14.8 kg/mol.

The difference between NMR and SEC measurements could be due to the calculation of the M_w and M_n with respect to a PI calibration. It is assumed that the structure of the diblock copolymer deviates from the structure of a PI homopolymer. Different arrangements of the polymer chains result in different hydrodynamic volumes [Lechner 03, Tieke 05] and, thus, the SEC calibration with PI is no longer valid for the diblock copolymer. In Table 8.1 the molecular weights determined by the different characterization techniques, SEC and NMR, are summarized for comparison and overview.

Table 8.1: Determined molecular weights by SEC with RI and IR detector and ^{1}H NMR analysis. The SEC-RI measurements are based on a PI calibration and SEC-IR on a PS calibration. The M_p of the PI homopolymer, measured via SEC-RI and SEC-IR, is doubled in comparison with the molecular weight of the PI precursor. In contrast to this the M_w of the PI homopolymer measured with the SEC-RI, has not doubled with respect to the M_w of the PI precursor. An explanation is the overlap of the PI homopolymer with the PI-b-PDMS peak in Figure 8.6 which generates lower molecular weight because the higher elution volumes are neglected.

determined struc-ture	M_n NMR (kg/mol)	M_n SEC-RI (kg/mol)	M_w SEC-RI (kg/mol)	M_p SEC-RI (kg/mol)	M_p SEC-IR (kg/mol)
PI-b-PDMS	14.8	10.6	11.0	10.9	10.1
PI precursor	/	3.1	3.3	3.3	2.8
PI homopolymer	/	4.5	4.8	4.9	5.6
PDMS block	11.7	/	/	/	/

8.5.3 DSC characterization

The differential scanning calorimetry (DSC) is used to have an indication for phase separation [Zhao 09]. A phase separated diblock copolymer shows two distinct glass transition temperatures T_g. The DSC is a thermoanalytical technique, which measures the changes in heat flux as a function of temperature, where it records the adsorption and delivery of heat of the sample to a reference [Wendlandt 97, Mathod 94]. DSC determines second order phase transition, like glass transition, as a step, whereas first order phase transitions as melting and crystallization appear as peaks in the DSC thermogram. Here, the DSC is measured with a Mettler Toledo DSC 30 in a temperature range from $-160\,°C$ to $25\,°C$. The heating rate is 2, 10 or $20\,K/min$ and nitrogen as gas atmosphere was used. The cis-1,4-PI, which is dominating in the synthesized PI-block, does not crystallize and thus possess only a glass transition temperature, T_g, at $-75\,°C$ [Cowie 97]. But the glass transition of the PI-block is covered by the incipient T_m peaks of the PDMS-block, Figure 8.10b. Generally it should be found where the square is placed. The baselines of the peaks 1 and 2, 3 gives an indication of a very weak step for T_g of PI (step-PI). In contrast, the

PDMS-block shows all well known phase transition processes. After the glass transition at $-125\,°$C, Figure 8.10, the PDMS is a fluid in a supercooled state [Cowie 97]. If temperature is raised, the polymer chains become more flexible and rearrange to form crystallites, which is measured in an exothermic crystallization peak at $T_c = -86\,°$C. This phenomenon is also called cold crystallization [Liu 00]. PDMS is a thermotropic liquid crystalline and therefore exhibits two endothermic melting peaks at $T_{m1} = -45\,°$C and $T_{m2} = -36\,°$C. In literature several possibilities are discussed, but it was found that the two melting peaks are due to the liquid crystalline domain structure [Liu 98, Liu 00, Schwarz 97]. Within a domain the chains are well aligned and crystallize easily. In the boundary region, the amorphous state with randomly oriented chains is present, which is hindered from crystallization. Therefore the crystallization is a slow process, which means that with a low heating rate peak 2 and peak 3 are well separated. Whereas at a high heating rate ($20\,$K/min), Figure 8.10c, peak 2 engages with peak 3 as it is shown by the shift of the maximum to higher temperatures from $2\,$K/min to $20\,$K/min. In contrast to this, the transition at T_{m2} belongs to a mesophase transition related to a fast process, which is not influenced by the heating rate [Liu 00].

For T_{m3} another melting peak is assumed, that belongs to a further isotropization process, where no obvious liquid crystalline structure was built [Liu 00].

Additionally it was shown that the glass transition for PI is not detected at the slowest heating rate. The increase to $10\,$K/min shows a weak indication for a glass transition of PI. A faster heating rate means less time for orientation of the polymer chains and, therefore, leads to a more pronounced glass transition. A further increase of the heating rate results in smearing of the crystallization peak of PDMS, which covers the glass transition of PI.

Since the DSC characterization yields an indication of two glass transition temperatures as an hint for phase separation, small angle X-ray scattering is used to determine the spatial distribution.

8.5.4 Two dimensional small angle X-ray scattering of the diblock copolymer PI-*b*-PDMS

Small angle X-ray scattering (SAXS) is used to characterize the structural properties of condensed matter on a length scale typically between 1 and $500\,$nm [Wedler 07]. Scattering

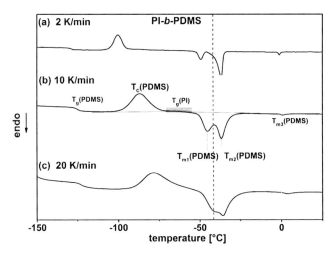

Figure 8.10: Measurement of a DSC of the diblock copolymer PI-b-PDMS in the temperature range from -160 to +25 °C with a heating rate of (a) 2 K/min, (b) 10 K/min and (c) 20 K/min.

in general is a physical process where radiation like light, sound or moving particles, are forced to deviate from their original straight trajectory by collisions with for example localized particles in the medium through which they pass. Scattering of an electromagnetic wave by a molecule or the electron shell of an atom induces for example an electrical dipole [Tipler 09]. This induced electrical dipole in turn irradiates again an electromagnetic wave in various spatial directions. The scattered electromagnetic wave of all scattering centers, like particles or atoms, interfere to the observed scattered intensity. Elastic scattering are called scattered electromagnetic waves with the same frequency like the incident irradiation. In inelastic scattering the incident irradiation frequency is decelerated while the scattering process. Generally within SAXS experiments only elastic scattering phenomena are observed [Kerker 69, Gerthsen 02, Wedler 07, Tipler 09]. Periodic structures with repetition units in the dimension of 1 to 100 nm (for the here used SAXS setup with a range for the scattering vector $\underline{q} = 0.06 - 4.7\,\mathrm{nm}^{-1}$) scatters irradiated X-ray with a specific angle, see Figure 8.11. This means the scattering angle gives information about the spatial arrangement of the system in the nm-scale. Bragg's equation defines the relation between spatial periodicity d, the irradiated wavelength and the scattering angle Θ under

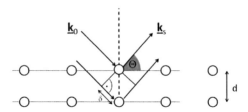

Figure 8.11: Principle of a scattering experiment with Θ being the scattering angle, and k_0 and k_s are the incident and scattered beam, respectively, $\delta = d \cdot \sin \Theta$ the extra length and d the spatial periodicity.

which constructive interference occurs:

$$n\lambda = 2d\sin(\Theta) \,. \tag{8.14}$$

Furthermore, $n\lambda$ measures the number of wavelength fitting between two rows of particles with distance d, thus, measuring reciprocal distances. From the vectorial description of scattering in Figure 8.11 the absolute values of the incident ($\underline{k}_0$) and irradiated ($\underline{k}_s$) wave vectors are defined as follows for an elastic scattering process where the irradiation frequency remains unchanged:

$$|\underline{k}_0| = |\underline{k}_s| = \frac{2\pi}{\lambda} \,. \tag{8.15}$$

From the wave vectors the scattering vector $\underline{q}$ is calculated which is in perpendicular direction to the scattering, see Figure 8.12:

$$\underline{q} = \underline{k}_0 - \underline{k}_s \,. \tag{8.16}$$

The absolute value of $|\underline{q}|$ is given by:

$$|\underline{q}| = \frac{4\pi}{\lambda}\sin(\Theta) \,. \tag{8.17}$$

From the combination of Equations (8.14) and (8.17) the correlation between the spatial distance d and the scattering vector is derived:

$$|\underline{q}| = \frac{2\pi}{d} \,. \tag{8.18}$$

The investigation of the diblock copolymer PI-b-PDMS via 2D SAXS gives information about the spatial heterogeneity induced by "microphase" separation of the immiscible

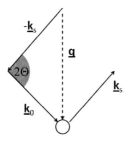

Figure 8.12: Schematic picture of the scattering vector $\underline{q}$ depending on the irratiated and scattered wave vectors $\underline{k}_0$ and $\underline{k}_s$.

blocks. The microphase separation is on the length scale of the diblock length. The length of a diblock copolymer is limited. The maximum is a totally stretched chain (here approximately 50 nm) and the minimum the conformation as a Gauss coil (here approximately 5 nm).

For diblock copolymers different microphase structures can be found. They depend on the Flory-Huggins interaction parameter χ_{FH}, the volume fraction of a block copolymer segment and the polymerization degree P_n. Figure 8.13a shows such a segregation diagram for a polystyrene-b-polyisoprene (PS-b-PI) diblock copolymer [Khandpur 95]. If the Flory Huggins interaction parameter is positive the polymer segments segregate due to a chemical diversity [Matsen 02, Thomas 88]. With a decreasing temperature χ_{FH} can be increased and, thus, the repulsive interaction between the two blocks increases and segregation occurs. This means, the larger χ_{FH} the stronger the tendency of segregation. As shown in Figure 8.13a also the polymerization degree has an influence on the microphase segregation. With an increasing polymerization degree raises the Flory-Huggins influence. The different structures are body centered cubic (BCC), hexagonal (H), gyroid (Q_{Ia3d}) and lamellar (L). An example for the different morphologies as a function of the volume fraction of the PS block (f_{PS}) is shown in Figure 8.13b for a PS-b-PI diblock copolymer with $\chi_{FH}P_n \approx 30$ [Khandpur 95]. For the here investigated diblock copolymer such a phase diagram has not been calculated, yet. Almdal et al. [Almdal 96] measured the χ_{FH} parameter for the diblock copolymer used here, PI-b-PDMS. Due to intensive experiments his research was attended to diblock copolymers with a volume fraction of the PDMS block $f_{PDMS} = 0.5$ and 0.65 and at measurement temperatures of 150 °C. Consequently the re-

Table 8.2: Characteristic reflexes of the different morphologies of diblock copolymers with a volume fraction of $f = 0.21$ of the minor block [Ulrich 00].

hexagonal	$\underline{q}_0$	$\sqrt{3}\underline{q}_0$	$\sqrt{4}\underline{q}_0$	$\sqrt{7}\underline{q}_0$	$\sqrt{8}\underline{q}_0$
body centered cubic	$\underline{q}_0$	$\sqrt{2}\underline{q}_0$	$\sqrt{3}\underline{q}_0$	$\sqrt{4}\underline{q}_0$	$\sqrt{5}\underline{q}_0$

sults cannot be used to predict a microphase symmetry for the here investigated diblock copolymer at room temperature with $f_{PI} = 0.21$. With the volume fraction of PDMS an estimation of the geometrical arrangement can be done with Figure 8.13a. With $f_{PI} = 0.21$ a hexagonal structure is expected for high polymerization degress P_n and a body centered cubic geometry for low P_n if the Flory-Huggins parameterχ_{FH} is assumed to be constant. Higher order reflexes are, therefore, expected at characteristic multiples of the first Bragg-reflex $\underline{q}_0$, see Table 8.2.

Different microphase symmetries give different reflections in $\underline{q}$ space [Hamley 98]. The different scattering reflections provide information about the spatial distribution of the scattering objects. Within this thesis a 2D SAXS experiment of the diblock copolymer PI-b-PDMS is measured, see Figure 8.14. The measurement shows a first order reflection $\underline{q}_0$ at $0.27\,nm^{-1}$. With Equation (8.18) the spatial separation of the immiscible blocks was determined to be 23 nm. The reflections at higher $\underline{q}$ values are too scattered in order to clearly identify them as higher orders of $\underline{q}_0$. Therefore, no conclusion about the microphase symmetry can be drawn. From the spatial separation of 23 nm it can be said, that the diblock copolymer chains are in between total stretching and a Gauss coil.

8.5.5 Conclusion of the PI-b-PDMS characterization

The combination of SEC with IR detection yields the measurement of the total molecular weight and chemical composition of the diblock copolymer. It was further used to determine the amount of PI homopolymer within the compatibilizer, which needed the measurement of a signal for PI as well as for PDMS. The detection of PDMS failed with all other available detectors like UV-, light scattering- and viscometer detector for the reasons mentioned in the beginning but was possible via SEC-IR. Thus, the newly invented measurement method extends the SEC application to polymers, which cannot be

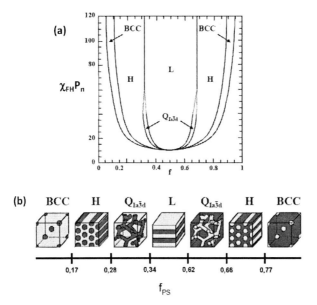

Figure 8.13: (a) Calculated phase diagram of the Flory-Huggins interaction parameter times the polymerization degree P_n for a polystyrene-b-polyisoprene (PS-b-PI) diblock copolymer as a function of the volume fraction of one block f [Khandpur 95]. (b) Calculated microphase symmetries for $\chi_{FH} P_n = 30$ with respect to the volume fraction of the PS block (=dark zones).

characterized with standard detectors. The DSC measurement has shown that spatial separation of the diblock copolymer occurs measured by two glass transition temperatures one for each block. The 2D SAXS scattering pattern could resolve the spatial heterogeneity but could not clearly resolve higher reflections of $\underline{q}_0$ and, thus, no information about the microphase symmetry could be obtained.

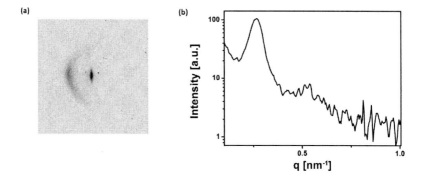

Figure 8.14: (a) 2D SAXS scattering pattern of PI-*b*-PDMS at 25 °C and ambient pressure. Due to the block collimator system only the half 2D SAXS pattern can be recorded. (b) Intensity of the 2D-SAXS pattern as a function of the scattering vector $\underline{q}$ after subtraction of the background. From the $\underline{q}$-vector at $0.27\,\mathrm{nm}^{-1}$ the spatial separation of the immiscible blocks of the diblock copolymer was determined to be 23 nm by Equation (8.18).

9. FT-Rheology on foams

9.1 Theory

As a last class of dispersions foams shall be introduced. Foams are a dispersion of a gas in a liquid or a solid continuous phase, whereas only liquids will be treated within this thesis. Foams are thermodynamically instable and can only be kinetically stabilized [Butt 10, Schuchmann 05]. Foams have a wide application range in daily life and industry. Beverages often contain carbondioxide as gas, like beer and softdrinks or air like in cappuccino. Foams based on milk protein stabilization can be found in several desserts like ice cream and mousse, to mention some examples [Schuchmann 05]. Adding yeast to dough or beer, carbondioxide is laid off during the production process and foam is generated. Of course foams are also used in industry due to their low density and low thermal conductivity. Styrofoam has a hollow structure and is used as packing and filling material. Metal foam with low density is applied in the automotive and aerospace industry [Butt 10].

A comparison of emulsions and foams yield major differences, which are summarized in the following based on [Schuchmann 05, Kraynik 88]:

- foams have not the low interfacial tension of generally $\Gamma < 5-10\,\mathrm{mN/m}$ for emulsions, but a rather high surface tension with approximately $30-50\,\mathrm{mN/m}$

- the density difference between dispersed phase and matrix is about $10^3\,\mathrm{kg/m^3}$ and, therefore, much higher than in most emulsions $(O(1\,\mathrm{kg/m^3}))$

- the viscosity ratio $\lambda = \frac{\eta_d}{\eta_m}$ is reasonably several decades lower than in the emulsions investigated here, for example between water and air $\lambda = \frac{0.02\,\mathrm{mPas}}{1\,\mathrm{mPas}} = 0.02$ [Sigloch 05]

- the bubbles possess a volumetric compressibility, in contrast to emulsions with water or oil as dispersed phase

- the bubble size is generally larger than $10\,\mu\mathrm{m}$ and thus one decade higher than in for example commercial emulsions with comparable volume fraction, see for example $\langle R \rangle_{43}$ of w/o-2 Section 7.7.1.

Foams are classified into three different categories depending on the amount of gaseous volume. A low volume fraction of dispersed phase, $\Phi_{\mathrm{Vol}} < 0.52$, is called a gas dispersion, where the bubbles appear as spheres without pronounced interactions [Manegold 53]. Volume fractions in between 0.52 and 0.72 are called "Kugelschaum" (German) or wet foam. The bubbles are spheres and densely packed until drainage and Ostwald ripening occur, see Section 9.2, and the volume fraction of the gas increases above 0.72. If Φ_{Vol} is higher than 0.72, the foam is also called dry foam or polyhedral foam [Butt 10]. The strucure of dry foam correlates with the stability. A metastable three dimensional foam follows some geometrical rules, which shall be mentioned [Butt 10]:

- in the polyhedra three flat sides meet at an angle of $120\,^\circ$

- no stability is reached, if four or more sides meet in one line

- at a corner, four edges meet at an polyhedral angle of $109.5\,^\circ$

- the smallest total film area should be reached, which is given for a polyhedra with 13.4 faces (experiments have found 14 and 12 faces as main and second choice, repectively).

9.2 Structure evolution in foams

The structure evolution of foam is mainly dominated by Ostwald ripening and drainage [Israelachvili 92]. Drainage is a fast process and accelerated by gravitational power. Generally the bubbles are not uniformly in size, but show a size distribution. The smaller the compartment, the higher the surface to volume ratio, which causes a transport of gas through the liquid layer from small bubbles into bigger bubbles, which is defined as Ostwald ripening. Therefore the bigger compartments grow on expense of the smaller ones.

The Laplace pressure between the compartments and the thin liquid layer is an additional driving force for Ostwald ripening. The pressure difference is defined by $\Delta p = 4\Gamma/R$ with R being the radius of curvature, see Figure 9.1. The prefactor 4 is given by the two sided planar film in a dry foam structure. Thus larger bubbles have a reduced pressure difference which enforces the Ostwald ripening process, see Figure 4.1. The order of magnitude of the Laplace pressure Δp in a bubble with $R = 100\,\mu m$ and $\Gamma = 40\,mN/m$ is $O(10^3\,Pa)$. Thus, the bubbles have a pressure of 1000 Pa higher than the liquid layer.

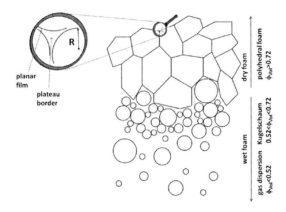

Figure 9.1: Different structures of a foam. With increasing volume fraction of the gaseous dispersed phase, the gas dispersion with single bubbles turns into a dry foam, where a densely packed structure of spheres goes into a polyhedral structure. The magnification shows a corner, where in a two dimensional demonstration three edges meet and build a plateau border. The plateau border deals as channel for the liquid when drainage takes place. The radius of curvature R determines the Laplace pressure $\Delta p = 4\Gamma/R$ between the plateau border and gaseous hollow [Butt 10].

Coarsening is an effect which takes some time, thus drainage is the predominant destabilization effect. The pressure in the plateau border is reduced in comparison to the gas phase, which allows the flow of the liquid into the plateau border. The plateau border is formed by the converge of the liquid films and acts as a channel. Within the planar films the hydrodynamic flow is slow, whereas once the liquid reaches the plateau border, the flow is additionally accelerated by gravitational forces to the downward direction. Thus drainage is the reason for drying of the foam at the top and wetting in the lower part. But drainage is not responsible for the growing of the bubbles it only affects the film thickness between the bubbles [Wierenga 05]. Figure 9.2a shows a photo of a beer foam during destabilization. The homogeneous foam with tiny bubbles is still visible in the lower foam part. The larger bubbles can be found at the top of the foam, due to drainage and Ostwald ripening effects. In Figure 9.2b the top view allows the detection of a beginning polyhedral structure, which will develop into a dry foam.

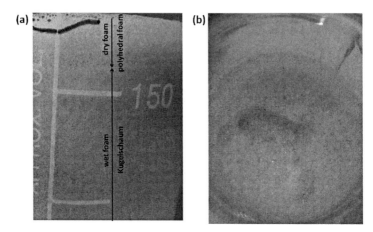

Figure 9.2: (a) Photo of a beer foam during its destabilization process. The homogeneous wet foam with small bubbles decreases due to drainage and Ostwald ripening. (b) After the half-life the foam has a changed structure showing the results of Ostwald ripening and drainage. The bubbles are larger and possess only a thin film layer. The foam is turning into a dry foam.

9.3 FT-Rheology applied on foams

FT-rheological investigation of foams is a specially demanding research with respect to the detection of nonlinearity due to the main constituents of water and air. Both have a low viscosity and water is additionally a Newtonian fluid. The Maffettone and Minale model, Equation (7.2), is the constitutive model for the nonlinear behavior of dilute emulsions, but does not consider dry foam properties like the low viscosity ratio and the high volume fraction generating interactions between the bubbles. Princen et al. [Princen 82, Princen 85] and Tcholakova et al. [Tcholakova 08] developed a rheological model based on the yield stress behavior of dry foams for the linear region and the steady state regime, see also Chapter 1. These models are not suitable to describe the large amplitude oscillatory shear behavior of the beer foam investigated within this thesis. But the simulations with the constitutive model of Maffettone and Minale in combination with the Batchelor theory have shown that the nonlinearity is influenced by the radius and interfacial tension of the droplets by $I_{5/3} \propto \frac{R^2}{\Gamma^2}$, Equation (7.4). Since the matrix itself is Newtonian, it is assumed to be not the source for nonliear behavior and, therefore, no further respected. Thus, the large bubble size and the high surface tension are assumed to mainly influence the nonlinear behavior of dry foam under LAOS. The measurement of nonlinearity requires a rheometer with high sensitivity, see Chapter 3. The stability of the foam is a limiting factor for the measurement time, thus the excitation frequency is chosen carefully to exploit the nonlinear behavior of a meaningful structure.

9.4 Formation of foam

Whipping and sparging are the two primary formation processes of foams. Both are combined within this work to achieve a homogeneous stable foam for LAOS investigations. Whipping incorporates air into the liquid beer state during for example rotation at high speed, whereas the sparging process rises gas bubbles into the beer [Wierenga 05]. A beaker with radius of 25 mm is screwed on top of a polypropylene funnel with a filter disc of Duran®. The pore size is $10 - 16\,\mu m$ as specified by the manufacturer, see Figure 9.3a. The funnel is connected with gaseous nitrogen, which sparges through the fresh beer of about 45 mL for each measurement. Nitrogen was chosen due to its decreased affinity to diffuse into the hydrophilic matrix in comparison with carbon dioxide. Nitrogen has a lower solubility in water than carbon dioxide with $1.7 \cdot 10^{-3}\,g/100g$ and $0.145\,g/100g$, respectively [Aylward 98]. If diffusion of the gas from the dispersed phase into the matrix is reduced, the stability of the foam can be prolonged. The beer is precooled in an icebath for about 1800 s to approximately 5 °C to increase stability. The gas flow was limited to a pressure of 0.5 Bar and applied for approximately 3 s in the beginning until the foam reaches a hight of about 100 mm. Simultaneously the whipping is started with a speed of 2000 rpm and lasts for 120 s to homogenize the generated bubbles. Within 60 s after formation of the foam the rheological experiments were applied.

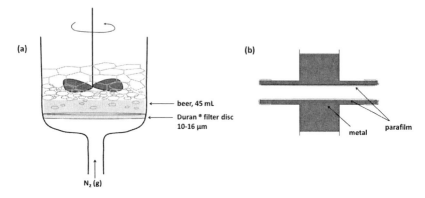

Figure 9.3: a) Setup for the formation of foam. Gaseous nitrogen sparges through the funnel and the filter disc of pore size $10 - 16\,\mu m$ into the $45\,mL$ fresh beer for $3\,s$. The electrical stirrer whips the foam with a speed of $2000\,rpm$ for $120\,s$. b) Covered $50\,mm$ plate plate geometry with parafilm to reduce the surface tension from above $O(1000\,mN/m)$ for metals to below $40\,mN/m$ for polymer surfaces [Israelachvili 92].

9.5 FT-Rheological experiments

The ARES G2 equipped with the FT software is used to measure the strain amplitude dependent nonlinear intensities. The measurement of $I_{n/1}$ of foams is a demanding task due to the low viscosity of the matrix. The high sensitivity of the ARES G2 was used to measure the relative third higher harmonic $I_{3/1}$. Higher nonlinearities are not available. Therefore, the intrinsic nonlinear parameter $^3Q_0 = Q_0$ was calculated from $I_{3/1}(\gamma_0)$ as introduced in Section 2.2.1 [Hyun 09].The nonlinear parameter Q_0 is a useful quantity to compare the nonlinear properties of different brands of beers.

The measurements were applied with a parallel plate of $50\,mm$ diameter to achieve a maximum torque resolution. A cone plate geometry was excluded because of the necessity of a defined gap, which is additionally too small for bubble sizes bigger than $0.1\,mm$ in diameter. The fresh prepared foam is loaded with a plastic spoon between the parallel plates, which were covered with a parafilm to reduce the surface tension and thus increase stability of the foam, see Figure 9.3b. The surface tension of a metal surface possess a surface tension of $O(1000\,mN/m)$ at room temperature, whereas a polymer surface tension, like the parafilm surface, is generally lower than $40\,mN/m$ [Israelachvili 92].

Within this thesis, six different brands of beer were investigated and their nonlinear rheological behavior compared. Guinness and Köstritzer were chosen as representatives of dark beer, whereas Rothaus Pils, Kölsch and Erdinger Hefe represent light beers. Kilkenny has a color which cannot be clearly allocated to one of these groups. But the FT-rheological experiments have shown, that it is related to the light beers, Figure 9.4. The measure-

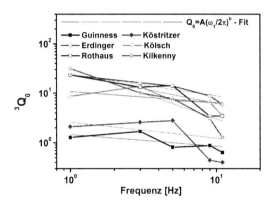

Figure 9.4: FT-rheological measurement of the nonlinear parameter Q_0 as introduced in Section 2.2.1. The Q_0 dependence on the excitation frequency shows a clear distinction between the nonlinear oscillatory shear behavior of dark and light beer. Within the Figure a power law fit, $Q_0(\omega_1/2\pi) = A \cdot (\omega_1/2\pi)^b$, results in different initial values A and nearly the same slopes b for all samples, which are summarized in Table 9.1.

ment of Q_0, Equation (2.12), as a function of frequency results in different curves for the six brands of beer. These spectra show decreasing Q_0 values with increasing frequency and are, therefore, fitted with a two parameter nonlinear fit comparable to an Ostwald de Waele fit for shear the shear thinning viscosity:

$$Q_0(\omega_1/2\pi) = A \cdot (\omega_1/2\pi)^b \qquad (9.1)$$

with b being the slope of the curves in the double logarithmic plot, Figure 9.4. The nonlinear fitting results in comparable slopes b but different values for A, which are summarized in Table 9.1. The parameter A, which is a measure for $I_{3/1}$ at small frequencies is for light beers one order of decade higher than for dark beers. The reason might be found in either a decreasing surface tension Γ or in an increasing radius R due the relation $I_{5/3} \propto \frac{R^2}{\Gamma^2}$

Table 9.1: Nonlinear fit parameters b and A from the nonlinear fit with the two parameter model $^3Q_0(\omega_1/2\pi) = A \cdot (\omega/2\pi)^b$ for the frequency dependent nonlinear mechanical parameter $^3Q_0$ for six different brands of beer. Guinness and Köstritzer are dark beers and the other four are light beers. The parameter A differentiate between the two types of beers by typically a factor of 10. Additionally the surface tension values are added

brand of beer	A	b	Γ (mN/m)
Guinness	1.46	-0.24	39.8
Köstritzer	2.61	-0.33	38.7
Erdinger	23.99	-0.43	36.4
Kölsch	10.82	-0.22	38.2
Rothaus Pils	24.15	-0.56	39.8
Kilkenny	31.58	-1.01	41.8

which is assumed to be the same for $I_{3/1} \propto \frac{R^2}{\Gamma^2}$, see Equation (7.4). Therefore the surface tension of the different beers were measured with the Wilhelmy plate in combination with the new ARES G2 setup introduced in Section 5.3.

9.6 Surface tension measurement of different brands of beer

The surface tension of beer foam is determined by the surface active substances like proteins [Roth 06]. The hydrophobic (or lipophilic) proteins concentrate at the inner surface of the bubbles due to repulsion with the hydrophilic matrix. The bittern in the matrix of the foam have unpolar side chains, which are attracted by the lipophilic proteins and settle, therefore, preferably at the lipophilic inner surface of the bubbles [Roth 06]. The surface tension of beer is measured with the Wilhelmy plate in combination with the ARES G2 rheometer as described in Section 5.3. The beer is cooled to a temperature of 10 °C. The Wilhelmy plate has room temperature like the ambient air. Reproducibility measurements result in the surface tensions for the different brands of beer listed in Table 9.1.

The surface tension values of the dark and light beers are within the range of 36.4 to

41.8 mN/m, whereby the dark and light beers do not show significant differences. The difference in nonlinearity, therefore, is attributed to the bubble size. The light beer has a nonlinearity of about a factor 10 higher than the dark beer determined by the parameter A from the nonlinear fit Equation (9.1). Due to Equation (7.4) the higher nonlinearity of light beer is assumed to be related to a larger bubble size since the surface tension is measured to be almost the same for dark and light beer. Thus, the nonlinearity is proposed to be a measure of the surface of the foam.

9.7 Outlook of beer research

Primarily experiments have shown that FT-Rheology is a sensitive measurement technique to measure the difference between dark and light beer. Although beer foam is a complex system with respect to the cereal grains used for fermentation and the proteins used for the stabilization of the foam, the nonlinear behavior is clearly divided into two groups of light and dark beer. The nonlinearity is expressed by the intrinsic nonlinear parameter Q_0 which is assumed to be influenced by the surface tension and the bubble size. It could be shown that the nonlinearity of light beer is about one decade higher than the nonlinearity of dark beer. As a possible explanation, this was further on referred to a larger bubble size in light beer because the surface tension measurements yield the same values for light and dark beer.

In future projects several parameters in beer foam preparation can be adjusted, like the gas for sparging or the rotational speed during whipping in order to achieve different bubble sizes.

In the rheological investigations the temperature effect could play an important role and is worth to examine. For sure these six brands of beer are a small excerpt from the large variety of beers and in future projects, more samples have to be measured. Guinness, for example, could show different behavior, if it is taken from glass bottles or stored in cans. From literature [Farrell 01, Son 83] it is known, that Guinness in glass bottles is sparged with carbondioxide, like all the other beers investigated above, whereas in cans it is sparged with nitrogen. Thus the foam of Guinness of cans could show different behavior from Guinness from glass bottles, although each is sparged with nitrogen in the experiments, because the initial condition is changed.

Other beverages with foams on top are for example cappuccino and latte macchiatto. Different proteins in milk and beer could generate different foam properties with respect to surface tension and bubble size.

It is obvious that beer foam investigation is a wide field of research and can be extended to many different directions. It will be of great interest how far FT-rheological experiments can be correlated to foam properties.

10. Conclusion

Fourier Transform Rheology was used to investigate materials with different space heterogeneities. The sensitivity of this measurement technique was determined by measuring the signal to noise ratio of the frequency dependent intensity spectrum. The first spectrum measured via nonlinear oscillatory rheology in Giacomin et al. [Giacomin 98] had a S/N of 10^2 and could be improved within this thesis about five decades to $S/N = 10^7$. To opimize sensitivity measurements with different types of rheometers have shown, that especially dispersions with a high volume fraction of the dispersed phase gain highly nonlinear behavior. The improvement in resolution of the torque detection down to $50\,\text{nNm}$ with the strain controlled rheometers led to the measurement of a new world record of detected higher harmonics in a frequency dependent spectrum with 289 overtones.

The main topic of this thesis is the characterization of emulsions via nonlinear measurements. It could be shown that simulations via constitutive models in combination with nonlinear oscillatory measurements exhibit emulsion properties like volume average radius and interfacial tension. The simulation parameters were varied to cover a wide range of emulsion properties like the radii $R = 0.1 - 100\,\mu\text{m}$ on a logarithmic scale, the excitation frequencies $\omega_1/2\pi = 0.1 - 5\,\text{Hz}$ on a linear scale, the matrix viscosities $\eta_m = 0.001 - 100\,\text{Pas}$ on a logarithmic scale, the viscosity ratios $\lambda = \frac{\eta_d}{\eta_m} = 2.5 - 10$ on a linear scale and the interfacial tensions $\Gamma = 2 - 50\,\text{mN/m}$ on a linear scale. Additionally the viscosity ratio dependent E_λ curves were simulated for different types of distributions $\Psi(R)$ like the Gaussian distribution, also called normal distribution, and the lognormal distribution, see

Section 7.6.1. For polydisperse emulsions a nonlinear mechanical master curve $^{n+2/n}E$ was established as a universal correlation among (i) the ratio of the higher harmonics $I_{n+2/n}$ with $n = 3$ and 5, (ii) the capillary number Ca and (iii) the viscosity ratio λ. In addition, it was shown that Ca must be based on the volume average radius $\langle R \rangle_{43}$ for $I_{5/3}$ and on $\langle R \rangle_{54}$ for $I_{7/5}$ to yield a master curve. At low strain amplitudes γ_0 the master curve has a constant value called nonlinear mechanical master number $^{n+2/n}E_0$ which is related to the emulsion properties and the rheological parameters by $\frac{^{n+2/n}Q_0}{\omega_1^2} = {^{n+2/n}}E^0\lambda^p\eta_m^2\langle R \rangle_{\frac{n+1}{2}+2;\frac{n+1}{2}+1}^2/\Gamma^2$. The nonlinear parameter $^{n+2/n}Q_0$ is determined by the scaling dependence of the relative higher harmonics on the strain amplitude with γ_0^{n-1}.

Additionally, a new methodology for the detection of the power law dependence of the $I_{n/1}$ components at low strain amplitudes γ_0 has been proposed: the overtones are approximated by a sum of two different scaling laws, $I_{n/1} = a_n \cdot \gamma_0^{-1} + b_n \cdot \gamma_0^{n-1}$ with n being an odd integer of one, which separate the region with γ_0^{-1} dependence dominated by the instrument noise at low γ_0 from the genuine power law scaling region with γ_0^{n-1} dependence.

However, for dilute emulsions, a state of the art rheometer with minimum measurable torque of $50\,\mathrm{nNm}$ still could not resolve the intrinsic strain amplitude dependence of $I_{7/1} \propto \gamma_0^6$ for emulsions with a low volume fraction, Equation (2.12). Consequently, at this point, dilute emulsions could not be analyzed so far using $^{7/5}E(Ca, \lambda)$.

In experiments PDMS in PIB and PDMS in PI blends were investigated with a volume fraction of $\Phi_{\mathrm{Vol}} = 10\,\%$ of the dispersed phase. The size distribution were adjusted by different homogenization processes and the interfacial tension by adding the corresponding commercially available diblock copolymer PIB-b-PDMS or the synthesized diblock copolymer PI-b-PDMS, respectively, as compatibilizer. The different polymer blends covers size distributions with volume average radii from $\langle R \rangle_{43} = 2\,\mathrm{\mu m}$ to $10\,\mathrm{\mu m}$ with polydispersities from $PDI = 1.17$ to 4.65 and interfacial tensions $\Gamma = 1.75\,\mathrm{mN/m}$ to $3.48\,\mathrm{mN/m}$. In contrast to this within the simulations the lower limit of the interfacial tension was determined to be $2\,\mathrm{mN/m}$.

The comparison of the characterization based on FT-rheological protocols with the Palierne model based on linear rheological measurements has shown, that the newly invented protocol is not limited to low polydispersities of $PDI = \frac{\langle R \rangle_{43}}{\langle R \rangle_{10}} < 1.8$ as the here applied Palierne model does. Additionally the nonlinear analyzing technique is not influenced by interfacial relaxation processes induced by the diblock copolymer in contrast to the influence on the Palierne model as applied within this thesis. The polymer blend PDMS/PI with diblock copolymer showed two relaxation processes which hindered the analysis via the Palierne model but still enabled a characterization via the nonlinear mechanical master number.

The E function was introduced for the characterization of dilute monodisperse emulsions. In particular, this analysis can be performed as well for dilute model dispersions, but also for highly concentrated commercial products with volume fractions Φ_{Vol} around 0.75. This is remarkable for FT-Rheolgoy as the traditional methods like scattering always required extremely high diluted systems, where dilution could influence the droplet size distribution. It has to be noted that the analysis via nonlinear rheology is a fast and an in-situ method. Commercial emulsions showed a significant amount of higher harmonics and could therefore be analyzed to determine their droplet size and width of distribution. Although theoretical effective models are not feasible for concentrated emulsions like for foam with high amount of dispersed phase, experimental evidences suggested that an empirical rule between volume average radius and non linear response can be however found. It could be shown that $\sqrt{5/3Q_0}$ is correlated to $\langle R \rangle_{43}$ and the ratio $\sqrt{7/5Q_0/5/3Q_0}$ exhibit information about the width of the distribution.

Beside the successful introduction of FT-Rheology in the field of emulsion characterization, this thesis offers the opportunity to look for other fields of characterization of dispersions where new information or new possibilities could be gained from the use of FT-Rheology. One successful example was the collaboration with Professor Fuchs (University Konstanz, Germany), Professor Brader (University Fribourg, Switzerland) and Professor Ballauff (Helmholtz Gemeinschaft Berlin, Germany). The well established and studied schematic $F_{12}^{(\dot{\gamma})}$ model based on the mode coupling theory was extended into the nonlinear regime. The measured polymer colloid with a volume fraction close to or beyond the glass transition, is assumed to behave in a first approximation like solid spheres. The Brownian motion of the particles is restricted to cages of adjacent particles. A basic consideration for calculating the nonlinear behavior was the yielding of the cage effect under nonlinear oscillatory shear. The aim of the project was the extending of the existing well-established schematic $F^{(\dot{\gamma})})_{12}$ model into the nonlinear regime. Linear measurements were conducted and used to determine the parameters for the simulation of the nonlinear behavior. For the first time the schematic microscopic approach was able to calculate the nonlinear oscillatory time behavior. The experimental evaluation was the last step in completing a new application area of MCT based theories. The overall good agreement between theory and experiments shows, that both enable a complete rheological characterization of colloidal systems up to the nonlinear regime and that further predictions via the MCT based theory can be assumed to be valid without additional experiments.

After investigation of solid and liquid dispersed particles in a liquid dispersed medium, nonlinear foam rheology were conducted. Foams have a gaseous dispersed phase and are only temporarily stable. Nonlinear rheological behavior is a demanding task since the viscosity of the matrix, mainly consisting of water, is very low. The high sensitivity of $S/N \approx 10^7$ of FT-rheological experiments was used to apply first experiments on foam. The experiments with FT-Rheology have shown that different brands of beers could be categorized via the intrinsic nonlinear ratio Q_0 into light and dark beer. Within this thesis the focus was on the principal possibility of FT-rheological measurements of foams. The frequency dependent Q_0 spectra were fitted with a two parameter model for monotonically decreasing curves $Q_0 = A(\omega_1/2\pi)^b$. The parameter A describes the nonlinearity at low frequencies and is for dark beer between $A = 1$ and 3 and for light beer between $A = 10$ and 30, thus one order of magnitude higher. According to the nonlinearity of emulsions, where the nonlinearity is amongst other parameters like the matrix viscosity and the rheological input parameters related to the radius and interfacial tension by $I_{5/3} \propto \frac{R^2}{\Gamma^2}$, the nonlinearity of foams was assumed to be influenced by mainly these two properties. The surface tension of the beer was measured with a combination of Wilhelmy plate and the axial force recording with the ARES G2, which is a new setup invented in this thesis for surface and interfacial tension measurements. The values of the surface tensions for the different brands of beer were around $40\,\mathrm{mN/m}$ and showed no significant difference between dark and light beer as they were differentiated by the Q_0 measurements. The difference in nonlinearity is, therefore, assumed to be based on different bubble sizes, where light beer seems to have larger bubbles than dark beer.

The applied surface measurement of water, PDMS and beer and interfacial tension measurements of polyethylene glycol dissolved in water against the oil miglyol with or without surfactant PGPR were applied by a combination of Wilhelmy plate and axial foce measurement with the ARES G2. The axial force measurement is limited to a minimum force of 0.001 N which yields a theoretical minimum surface and interfacial tension of $\Gamma = 25\,\text{mN/m}$. But in experiments it could be shown that an interfacial tension down to $5\,\text{mN/m}$ could still be resolved.

It can be presumed that FT-Rheology offers a wide field of application in dispersion characterization. A combination of theory and experiment was proven several times within this thesis to achieve an advanced characterization of dispersions. Improvements in mathematical simulation procedures as well as in the hardware of the rheometers opens new research areas, as shown by the different examples in this thesis. FT-Rheology can be used as a non-invasive characterization tool for model emulsions, but it offers also the potential of in-situ characterization of highly concentrated and highly complex commercial systems like w/o-emlsions and beer foams.

10.1 Outlook

Investigation of the nonlinear behavior of dilute emulsions consisting of Newtonian droplets in a Newtonian matrix is based on the droplet deformation under LAOS. With this work it should be shown, that the detected nonlinear behavior is a useful tool for characterizing emulsion properties. The challenging task of this analyzing method is the detection of the intrinsic nonlinear behavior at small strain amplitudes. The improvement in sensitivity was therefore a development, which has improved the newly invented characterization method. For example the intensity of the seventh overtone could not be achieved via the ARES G1 and ARES G2 for dilute polymer blends like PDMS/PIB and PDMS/PI. Future prospects will be the measurement of dilute emulsions with the Hybrid Rheometer DHR2. The lower torque limit of $2\,\text{nNm}$ of the DHR2 instead of $50\,\text{nNm}$ of the ARES G2 should exhibit the intrinsic behavior of $I_{7/1}$ as a function of strain amplitude. Then, an access to the width of the droplet distribution is also achieved for dilute emulsions.

To increase the sensitivity of surface and interfacial tension measurements with the Wilhelmy plate, the minimum axial force has to be lowered. Within the group of Professor Wilhelm the Capillary breakup elongational rheometer (CaBER) was provided with highly sensitive transducers [Klein 09], which enable measurements of axial forces down to 10^{-5}N

which yields a theoretical minimum surface or interfacial tension of $0.2\,\mathrm{mN/m}$. Thus, a combination of the CaBER and the Wilhelmy plate could improve the interfacial and surface tension measurements with respect to the minimum measurable Γ and accuracy of Γ.

In the case of foam rheology the first measurements exhibit interesting results about the classification of beer foam. Future work will focus on a correlation parameter like the bubble size for the different nonlinear behavior. As well the types of foams will be extended to a wider field of daily life products. For example milk foam, as it is found on several coffee beverages, should be worth the investigation. Additionally, the preparation of foam with different gas and homogenization processes is of interest.

Bibliography

[Almdal 96]   K. Almdal, K. Mortensen, A.J. Ryan, F.S. Bates. *Macromolecules* **29**, 5940 (1996).

[Almusallam 04]   A.S. Almusallam, R. Larson, M.J. Solomon. *Polymers* **48**, 319 (2004).

[Angell 94]   C.A. Angell, P. H. Poole, J. Shao. *Nuovo Cimento D* **16**, 993 (1994).

[Atkins 07]   P.W. Atkins, J. de Paula. Physikalische Chemie. Wiley-VCH, Weinheim (2007).

[Aylward 98]   G.H. Aylward, T.J.V. Findlay. Datensammlung Chemie in SI-Einheiten, Volume 3. Wiley-VCH, Weinheim (1998).

[Batchelor 70]   G. Batchelor. *J. Fluid Mech.* **41**, 545 (1970).

[Belitz 08]   H.D. Belitz, W. Grosch, P. Schieberle. Das Lehrbuch der Lebensmittelchemie. Springer Verlag, Berlin (2008).

[Bellas 00]   V. Bellas, H. Iatrou, N. Hadjichristidis. *Macromolecules* **33**, 6993 (2000).

[Beskers 10]   T.F. Beskers. On-line Kopplung und Optimierung der FT-IR-Spektroskopie mit der Gelpermeationschromatographie zur Polymercharakterisierung. Diploma thesis, Karlsruhe Institute of Technology (2010).

[Beskers 11]   T.F. Beskers. *submitted* (2011).

[Bird 87]   R.B. Bird, R.C. Armstrong, O. Hassager. Dynamics of Polymeric Liquids. Wiley, New York (1987).

[Bousmina 00]   M. Bousmina, M.B. Aouina, R. Chaudhry, R. Guenette, R.E.S. Bretas. *Rheol. Acta* **40**, 538 (2000).

[Bracewell 86]   R.N. Bracewell. The Fourier Transform and its Application, Volume 2nd edition. McGraw-Hill, New York (1986).

[Brader 07]   J.M. Brader, Th. Voigtmann, M. E. Cates. *Phys. Rev. Lett.* **98**, 058301 (2007).

[Brader 08]   J.M. Brader, M. E. Cates, M. Fuchs. *Phys. Rev. Lett.* **101**, 138301 (2008).

[Brader 09]   J.M. Brader, T. Voigtmann, M. Fuchs, R. G. Larson, M. E. Cates. *Proc. Natl. Acad. Sci. U.S.A.* **106**, 15186 (2009).

[Brader 10]   J. M. Brader, M. Siebenbürger, M. Ballauff, K. Reinheimer, M. Wilhelm, S.J. Frey, F. Weysser, M. Fuchs. *Phys. Rev. E* **82**, 0614011 (2010).

[Brandrup 99]   J. Brandrup, E.H. Immergut, E.A. Grulke. Polymer Handbook. Wiley, New Jersey (1999).

[Bronstein 08]   I.N. Bronstein, K.A. Semendjajew, G. Musiol, H. Mühlig. Taschenbuch der Mathematik. Verlag Harri Deutsch (2008).

[Brummer 06]   R. Brummer. Rheology Essentials of Cosmetic and Food emulsions. Springer (2006).

[Burlando 10]   B. Burlando, L. Verotta, L. Cornara, E. Bottini-Massa. Herbal principles in cosmetics. CRC Press Taylor and Francis Group (2010).

[Butt 10]   H.J. Butt, K. Graf, M. Kappl. Physics and Chemistry of Interfaces. Wiley-VCH, Weinheim (2010).

[Butz 98]   T. Butz. Fourier Transformation für Fußgänger. Teubner, Stuttgart, Leipzig (1998).

[Carotenuto 08]   C. Carotenuto, M. Grosso, P.L. Maffettone. *Macromolecules* **41**, 4492 (2008).

[Chesters 91]   A.K. Chesters. *Trans. I Chem. E* **69**, 259 (1991).

[Cho 05]   K.A. Cho, K. Hyun, K.H. Ahn, S.J. Lee. *J. Rheol.* **49**, 747 (2005).

[Claridge 99]   T.D.W. Claridge. High-Resolution NMR Techniques in Chemistry Research. Pergamon Press, Amsterdam (1999).

[Clemens 04]   D.J. Mc Clemens. Food, Emulsions, Principles, Practices and Techniques. CRC Press, University of Massachusetts, Amherst (2004).

[Consul 73]   P.C. Consul, G.C. Jain. *Technometrics* **15**, 791 (1973).

[Cooley 65]   J.W. Cooley, J.W. Tuckey. *Math. Comput.* **19**, 297 (1965).

[Cosgrove 05]   T. Cosgrove. Colloid Science, Principles, Methods and Applications. Blackwell Publishing, Oxford (2005).

[Cowie 97]   J.M.G. Cowie. Chemie und Physik der synthetischen Polymeren. Vieweg, Braunschweig (1997).

[Crassous 06]   J.J. Crassous, M. Ballauff, M. Drechsler, J. Schmidt, Y. Talmon. *Langmuir* **22**, 2403 (2006).

[Crassous 08a]   J.J. Crassous, M. Siebenbürger, M. Ballauff, M. Drechsler, D. Hajnal, O. Henrich, M. Fuchs. *J. Chem. Phys.* **128**, 204902 (2008).

[Crassous 08b]   J.J. Crassous, A. Wittemann, M. Siebenbürger, M. Schrimmer, M. Drechsler, M. Ballauff. *Colloid Polym. Sci.* **286**, 805 (2008).

[Das 05]   N.C. Das, H. Wang, J. Mewi, P. Moldenaers. *J. Pol. Sci. Part B Polymer Physics* **43**, 3519 (2005).

[Dealy 06]   J.M. Dealy, R.G. Larson. Structure and Rheology of molten polymers: From Structure to Flow behavior and back again. Hanser Publishers, Munich (2006).

[Denkov 09]   N.D. Denkov, S. Tcholakova, K. Golemanov, K.P. Ananthpadmanabhan, A. Lips. *Soft Matter* **5**, 3389 (2009).

[Derkach 09]   S.R. Derkach. *Adv. Colloid Interface* **151**, 1 (2009).

[Dingenouts 10]   N. Dingenouts, A. Horvath. Polymer-Praktikum: Grundlagen, Versuchsbeschreibungen. Karlsruhe (2010).

[Dörfler 02]   H.D. Dörfler. Grenzflächen und kolloid-disperse Systeme. Springer, Berlin (2002).

[Draper 98]   N.R. Draper, H. Smith. Applied Regression Analysis. Wiley and Sons, New York (1998).

[Elkins 04]   C.L. Elkins, T.E. Long. *Macromolecules* **37**, 6657 (2004).

[Ewoldt 08]   R.H. Ewoldt, A.E. Hosoi, G.H. McKinley. *J. Rheol.* **52**, 1427 (2008).

[Fahrländer 99]   M. Fahrländer, C. Friedrich. *Rheol. Acta* **38**, 1 (1999).

[Farrell 01]   S. Farrell, R.P. Hesketh, J. A. Newell, C.S. Slater. *Int. J. Engng. Ed.* **17**, 588 (2001).

[Fearn 99]   T. Fearn. *Spectr. Europe* **11**, 24 (1999).

Bibliography

[Fischer 91] K.H. Fischer. Physik der Polymere: Der Glasübergang nach der Modenkopplungstheorie. Forschungszentrum Jülich (1991).

[Flory 53] P.J. Flory. Principles of Polymer Chemistry. Cornell University Press, New York (1953).

[Franck 08] A. Franck. *Appl. Rheol.* **18**, 44 (2008).

[Friedrich 95] C. Friedrich, W. Gleisner, E. Korat, D. Maier, J. Weese. *J. Rheol.* **39**, 1411 (1995).

[Fuchs 95] M. Fuchs. *Transport Theor. Stat.* **24**, 855 (1995).

[Fuchs 02] M. Fuchs, M.E. Cates. *Phys. Rev. Lett.* **89**, 248304 (2002).

[Fuchs 03] M. Fuchs, M.E. Cates. *Faraday Discuss.* **123**, 267 (2003).

[Fuchs 09] M. Fuchs, M.E. Cates. *J. Rheol.* **53**, 957 (2009).

[Fulcher 25] G.S. Fulcher. *J. Am. Ceram. Soc.* **8**, 339 (1925).

[Fuoss 54] R.M. Fuoss. *J. Am. Chem. Soc.* **76**, 5905 (1954).

[Gerthsen 02] C. Gerthsen, D. Meschede, H. Vogel. Physik: die ganze Physik zum 21. Jahrhundert. Springer, Berlin (2002).

[Giacomin 93a] A.J. Giacomin, R.S. Jeyaseelan, T. Samurkas, J.M. Dealy. *J. Rheol.* **37**, 811 (1993).

[Giacomin 93b] A.J. Giacomin, J.G. Oakley. *Rheol. Acta* **32**, 328 (1993).

[Giacomin 98] A.J. Giacomin, J.M. Dealy. Using Large-Amplitude Oscillatory Shear. A.A. Collyer, D.W. Clegg, (eds.), Chapman and Hall, London (1998).

[Götze 91] W. Götze, J.P. Hansen, D. Levesque, J. Zinn-Justin. Aspects of structural glass transitions in Liquids, Freezing and Glass Transition. Les Houches Summer Schools of Theoretical Physics, Norh-Holland, Amsterdam (1991).

[Götze 92] W. Götze, L. Sjogren. *Rep. Prog. Phys.* **55**, 241 (1992).

[Grace 82] H.P. Grace. *Chem. Eng. Comm.* **14**, 225 (1982).

[Graebling 93a] D. Graebling, R. Muller, J.F. Palierne. *J. Phys. IV* **3**, 1525 (1993).

[Graebling 93b] D. Graebling, R. Muller, J.F. Palierne. *Macromolecules* **26**, 320 (1993).

[Green 54] M.S. Green. *J. Chem. Phys.* **22**, 398 (1954).

[Grosso 07]   M. Grosso, P.L. Maffettone. *J. Non-Newtonian Fluid* **48**, 143 (2007).

[Guido 00]   S. Guido, M. Minale, P.L. Maffettone. *J. Rheol.* **44**, 1385 (2000).

[Guido 04]   S. Guido, M. Grosso, P.L. Maffettone. *Rheol. Acta* **43**, 575 (2004).

[Guschl 03]   P.C. Guschl, J.U. Otaigbe. *J. Colloid Interface Sci.* **266**, 82 (2003).

[Hadjichristidis 00] N. Hadjichristidis, H. Iatrou, S. Pispas, M. Pitsikalis. *J. Polym. Sci. Part A: Polym. Chem.* **38**, 3211 (2000).

[Hamley 98]   I.W. Hamley. The physics of block copolymers. Oxford University Press, Oxford (1998).

[Helfand 82]   E. Helfand, D.S. Pearson. *J. Polym. Sci. Polym. Phys.* **5**, 217 (1982).

[Heuer 95]   H. Heuer, M. Wilhelm, H. Zimmermann, H.W. Spiess. *Phys. Rev. Letter* **75**, 2851 (1995).

[Higgins 76]   R.J. Higgins. *Am. J. Phys.* **44**, 766 (1976).

[Hilliou 04]   L. Hilliou, D. van Dusschoten, M. Wilhelm, H. Burhin, E.R. Rodger. *Rubber Chem. Technol.* **77**, 192 (2004).

[Homans 89]   S.W. Homans. A Dictionary of Concepts in NMR. Clarendon Press, Oxford (1989).

[Honerkamp 94]   J. Honerkamp. Stochastic Dynamical Sytems: Conceps, Numerical Methods, Data Analysis. VCH, New York (1994).

[Howe 97]   J.M. Howe. Interfaces in Materials. Wiley Interscience, New York (1997).

[Hsieh 96]   H.L. Hsieh, R.P. Quirk. Anionic polymerization: Principles and practical applications. Marcel Dekker, New York (1996).

[Huitric 49]   J. Huitric, P. Mederic, M. Moan, J. Jarrin. *Polymer* **39**, 1998 (4849).

[Hyun 03]   K. Hyun, J.G. Nam, M. Wilhelmy, K.H. Ahn, S.J. Lee. *Australia Rheol. J.* **15**, 97 (2003).

[Hyun 07]   K. Hyun, E.S. Baik, K.H. Ahn, S. Jong Lee. *J. Rheol.* **51**, 1319 (2007).

[Hyun 09]   K. Hyun, M. Wilhelm. *Macromolecules* **42**, 411 (2009).

[Hyun 11]   K. Hyun, M. Wilhelm, C.O. Klein, K. Soo Cho, J. Gun Nam, K. Hyun Ahn, S. Jong Lee, R.H. Ewoldt, G.H. Kinley. *Prog. Polym. Sci.* **36**, 1697 (2011).

[Israelachvili 92]   J. Israelachvili. Intermolecular and Surface forces. Academic Press, London (1992).

[Jackson 03]   N.E. Jackson, C.L. Tucker. *J. Rheol.* **47**, 659 (2003).

[Jacobs 99]   U. Jacobs, M. Fahrländer, J. Winterhalter, C. Friedrich. *J. Rheol.* **43**, 1495 (1999).

[Jansseune 00]   T. Jansseune, J. Mewis, P. Moldenaers, M. Minale, P.L. Maffettone. *J. Non-Newtonian Fluid* **93**, 153 (2000).

[Jeon 03]   H.K. Jeon, C.W. Macosko. *Polymer* **44**, 5381 (2003).

[Jondral 02]   F. Jondral, A. Wiesler. Wahrscheinlichkeitsrechnung und stochastische Prozesse. Teubner (2002).

[Kallus 01]   S. Kallus, N. Willenbacher, S. Kirsch, D. Distler, T. Neidhöfer, M. Wilhelm, H.W. Spiess. *Rheol. Acta* **40**, 552 (2001).

[Kammeyer 02]   K.D. Kammeyer, K. Kroschel. Digital Signalverarbeitung. Filterung und Spektralanalyse mit MATLAB-Übungen. Teubner (2002).

[Kennedy 94]   M.R. Kennedy, C. Pozrikidis, R. Skalak. *Comput. Fluids* **23**, 251 (1994).

[Kerker 69]   M. Kerker. The scattering of light and other electromagnetic radiation. Acad. Press, New York (1969).

[Khandpur 95]   A.K. Khandpur, S. Forster, F.S. Bates, I.W. Hamley, A.J. Ryan, W. Bras, K. Almdal. *Macromolecules* **28**, 8796 (1995).

[Kitade 97]   S. Kitade, A. Ichikawa, N. Imura, Y. Takahashi, I. Noda. *J. Rheol.* **41**, 1039 (1997).

[Klein 05]   C. Klein. Rheology and Fourier-Transform Rheology on water-based systems. *Ph.D. thesis*, Johannes Gutenberg-Universität Mainz (2005).

[Klein 07]   C. Klein, H.W. Spiess, A. Callin. *Macromolecules* **40**, 4250 (2007).

[Klein 09]   C. Klein, I.F.C. Naue, J. Nijman, H. Buggisch, M. Wilhelm. *Soft Materials* **7**, 242 (2009).

[Kob 02]   W. Kob. *Les Houches Summer School* **Session LXXVII** (2002).

[Koberstein 90]   J.T. Koberstein. *Wiley, New York* (1990).

[Kraynik 88]   A.M. Kraynik. *Ann. Rev. Fluid Mech.* **20**, 325 (1988).

[Kubo 57]	R. Kubo. *J. Phys. Soc. Jpn.* **12**, 570 (1957).
[Kuraray 11]	Kuraray, Liquid Rubber. website (2011-22-11). `http://www.kuraray.co.jp/en/products/chemical/pdf/lir_catalog.pdf`.
[Lacroix 96]	C. Lacroix, M. Bousmina, P.J. Carreau, B.D. Favis, A. Michel. *Polymer* **37**, 2939 (1996).
[Larson 88]	R.G. Larson. Constitutive equation for polymer melts and solutions. Butterworths, London (1988).
[Larson 99]	R.G. Larson. The structure and Rheology of Complex Fluids. Oxford University Press, Oxford (1999).
[Lechner 03]	M.D. Lechner, K. Gehrke, E.H. Nordmeier. Makromolekulare Chemie. Birkhäuser (2003).
[Li 97]	X. Li, C. Pozrikidis. *J. Fluid Mech.* **341**, 165 (1997).
[Liu 98]	S.L. Liu, T.S. Chung, S.H. Goh, Y. Torii, A. Yamaguchi, Ohta. *M. Polym. Eng. Sci.* **38**, 1845 (1998).
[Liu 00]	S.L. Liu, T.S. Chung, H. Oikawa, A. Yamaguchi. *Journal of Polymer Science Part B: Polymer Physics* **38**, 3018 (2000).
[M. Siebenbürger 06]	M. Siebenbürger. *Diploma thesis, Universität Bayreuth* (2006).
[Macosko 94]	C.W. Macosko. Rheology: Principles, Measurements and Application. VCH publishers, Inc., New York (1994).
[Maffettone 98]	P.L. Maffettone, M. Minale. *J. Non-Newtonian Fluid* **78**, 227 (1998).
[Manegold 53]	E. Manegold. Schaum. Chemie und Technik Verlagsgesellschaft, Heidelberg (1953).
[Maschke 92]	U. Maschke, T. Wagner. *Makromol. Chem.* **193**, 2453 (1992).
[Mathod 94]	V.B.F. Mathod. Calorimetry and Thermal Analysis of Polymers. C. Hanser, Munich (1994).
[Matsen 02]	M.W. Matsen. *J. Phys.: Condens. Matter* **14**, 21 (2002).
[Meins 11]	T. Meins, K. Hyun, N. Dingenouts, M. Fotouhi Ardakani, B. Struth, M. Wilhelm. *Macromolecules*, doi.org/10.1021/ma201492n (2011).
[Mezger 06]	T.G. Mezger. The Rheology Handbook. Vincentz, Hannover (2006).
[Milner 96]	S.T. Milner, H.W. Xi. *J. Rheol.* **40**, 663 (1996).

[Mori 99]   S. Mori, H.G. Barth. Size Exclusion Chromatography. Springer (1999).

[Müller 09]   A.H.E. Müller, K. Matyjaszewski. Controlled and living polymeriza-
                 tions: methods and materials. Wiley-VCH, Weinheim (2009).

[Nakayama 99]   Y. Nakayama, R.F. Boucher. Introduction to Fluid Mechanics.
                 Butterworth-Heinemann, Oxford (1999).

[Napper 83]   D.H. Napper. Polymeric Stabilization of Colloidal Dispersions. Aca-
                 demic Press, London (1983).

[Neidhöfer 03]   T. Neidhöfer. Fourier-transform rheology on anionically synthesised
                 polymer melts and solutions of various topology. *Ph.D. thesis*, Jo-
                 hannes Gutenberg-Universität Mainz (2003).

[Neidhöfer 04]   T. Neidhöfer, S. Siuola, N. Hadjichristidis, M. Wilhelm. *Macromol.
                 Rapid Commun.* **25**, 1921 (2004).

[Nicomp manual 11] Nicomp manual. website (2011-22-11). `http://http:`
                 `//www.colorado.edu/ceae/environmental/ryan/research/pdfs/`
                 `pss-380-manual.pdf`.

[Odian 04]   G.G. Odian. Principles of polymerization. Wiley-Interscience, Hoboken
                 N.J. (2004).

[Ottino 99]   J.M. Ottino, P. De Roussel, S. Hansen, D.V. Khakhar. *Adv. Chem.
                 Eng.* **25**, 105 (1999).

[Pal 06]   R. Pal. Rheology of Particulate Dispersions and Composites. CRC
                 Press (2006).

[Palierne 90]   J.F. Palierne. *Rheol. Acta* **29**, 204 (1990).

[Palierne 91]   J.F. Palierne. *Rheol. Acta* **30**, 497 (1991).

[Pearson 82]   R.S. Pearson, W.E. Rochefort. *J. Polym. Sci. Polym. Phys.* **20**, 83
                 (1982).

[Piirma 92]   I. Piirma. Polymeric Surfactants. Dekker, New York (1992).

[Pipkin 72]   A.C. Pipkin. Lectures on viscoelastic theory. Springer-Verlag, New
                 York (1972).

[Poon 95]   W.C.K. Poon, P.N. Pusey. Observation, Prediction and Simulation of
                 Phase Transitions in Complex Fluids. Kluwer Academic Publishers
                 (1995).

[Princen 82]	H.M. Princen. *J. Colloid Interf. Sci.* **91**, 160 (1982).
[Princen 85]	H.M. Princen. *J. Colloid Interf. Sci.* **75**, 246 (1985).
[Princen 86]	H.M. Princen, A.D. Kiss. *J. Colloid Interf. Sci.* **112**, 427 (1986).
[Princen 01]	H.M. Princen. The Structure, Mechanics, and Rheology of Concentrated Emulsions and Fluid Foams, Encyclopedia of Emulsions Technology. Dekker, New York (2001).
[Rallison 84]	J.M. Rallison. *Annu. Rev. Fluid Mech.* **16**, 45 (1984).
[Ramirez 95]	R.W. Ramirez. The FFT Fundamentals and Concepts. Engelwood Cliffs, Prentice-Hall (1995).
[Reinheimer 11a]	K. Reinheimer, M. Grosso, F. Hetzel, J. Kübel, M. Wilhelm. Fourier Transform Rheology as a noninvasive morphological characterization technique for the emulsion volume averge radius and its distribution. *J. Colloid Interf. Sci.*, submitted (2011).
[Reinheimer 11b]	K. Reinheimer, M. Grosso, M. Wilhelm. *J. Colloid Interf. Sci.* **360**, 818 (2011).
[Roth 06]	K. Roth. *Chem. Unserer Zeit* **40**, 338 (2006).
[Rusu 99]	D. Rusu, E. Peuvrel-Disdier. *J. Rheol.* **43**, 1391 (1999).
[Sasol Germany GmbH 11]	Sasol Germany GmbH. website (2011). `http://abstracts.aapspharmaceutica.com/ExpoAAPS05/Data/EC/Event/Exhibitors/312/8f67af20-407b-49ab-bd8c-9de42fbfa235.pdf`.
[Schlatter 05]	G. Schlatter, G. Fleury, R. Muller. *Macromolecules* **38**, 6492 (2005).
[Schmidt-Rohr 94]	K. Schmidt-Rohr, H.W. Spiess. Multidimensional solid state NMR and Polymers. Academic Press (1994).
[Schuchmann 05]	H.P. Schuchmann, H. Schuchmann. Lebensmittelverfahrenstechnik, Rohstoffe, Prozesse, Produkte. Wiley VCH, Weinheim (2005).
[Schwarz 97]	G. Schwarz, S.J. Sun, H.R. Kricheldorf. *Macromol. Chem. Phys.* **198**, 3123 (1997).
[Siebenbürger 09]	M. Siebenbürger, M. Fuchs, H. Winter, M. Ballauff. *J. Rheol.* **53**, 707 (2009).
[Siebenbürger 10]	M. Siebenbürger. Rheology of Concentrated Colloidal Suspensions and Comparison with Mode Coupling Theory. *Ph.D. thesis*, Universität Bayreuth (2010).

[Sigloch 05]   H. Sigloch. Technische Fluidmechanik. Springer, Berlin (2005).

[Skoog 92]   D. A. Skoog, J. J. Leary. Principle of Instrumental Analysis. Harcourt Brace College Publishers, Orlando (1992).

[Son 83]   A. Guinness Son, Ltd. Company. Beverage Package and a Method of Packaging a Beverage Containing Gas in a Solution. U.S. Patent number 4,832,968 (1983).

[Stieß 09]   M. Stieß. Mechanische Verfahrenstechnik und Partikeltechnologie. Springer, Berlin (2009).

[Stone 89]   H.A. Stone, L.G. Leal. *J. Fluid Mech.* **198**, 399 (1989).

[Stone 90]   H.A. Stone, L.G. Leal. *J. Fluid Mech.* **220**, 161 (1990).

[Szwarc 56]   M. Szwarc. *Nature* **178**, 1168 (1956).

[Szwarc 69]   M. Szwarc. Carbanions, Living Polymers and Electron Processes. Interscience Publishers (1969).

[TA Instruments 11] TA Instruments. website (2011-22-11). `http://www.tainstruments.com/product.aspx?siteid=11&id=257&n=1`.

[Tcholakova 08]   S. Tcholakova, N.D. Denkov, K. Golemanov, K.P. Ananthapadmanabhan, A. Lips. *Phys. Rev. E* **78**, 011405 (2008).

[Tee 75]   T.T. Tee, J.M. Dealy. *Trans. Soc. Rheol.* **19**, 595 (1975).

[Thomas 88]   E.L. Thomas, D.M. Anderson, C.S. Henkee, D. Hoffman. *Nature* **334**, 598 (1988).

[Tieke 05]   B. Tieke. Makromolekulare Chemie: Eine Einführung. Wiley-VCH, Weinheim (2005).

[Tipler 09]   P.A. Tipler, G. Mosca. Physik für Wissenschaftler und Ingenieure. Spektrum, Heidelberg (2009).

[Tracht 98]   U. Tracht, M. Wilhelm, A. Heuer, H. Feng, K. Schmidt-Rohr, H.W. Spiess. *Phys. Rev. Letter* **81**, 2727 (1998).

[Tucker 02]   C.L. Tucker, P. Moldenaers. *Annu. Rev. Fluid Mech.* **34**, 177 (2002).

[Ulrich 00]   R. Ulrich. Morphologie und Eigenschaften strukturierter organisch-anorganischer Hybridmaterialien. PhD thesis, Johannes Gutenber-Universität Mainz (2000).

[van Dusschoten 01]   D. van Dusschoten, M. Wilhelm. *Rheol. Acta* **40**, 395 (2001).

[van Hemelrijck 04]   E. van Hemelrijck, P. van Pyuvelde, C.W. Macosko, P. Moldenaers. *J. Rheol.* **48**, 143 (2004).

[van Hemelrijck 05]   E. van Hemelrijck, P. van Pyuvelde, C.W. Macosko, P. Moldenaers. *J. Rheol.* **49**, 783 (2005).

[van Megen 94]   W. van Megen, S. M. Underwood. *Phys. Rev. E* **49**, 4206 (1994).

[van Pyuvelde 02]   P. van Pyuvelde, S. Velankar, J. Mewis, P. Moldenaers. *Polym. Eng. Sci.* **42**, 1957 (2002).

[Velankar 01]   S. Velankar, P. van Pyuvelde, J. Mewis, P. Moldenaers. *J. Rheol.* **45**, 1007 (2001).

[Vincze-Minya 07]   K.A. Vincze-Minya, A. Schausberger. *J. Appl. Polym. Sci.* **105**, 2294 (2007).

[Vogel 21]   H. Vogel. *Phys. Z.* **22**, 645 (1921).

[Watanabe 11]   H. Watanabe, Q. Chen, Y. Kawasaki, Y. Matsumiya. *Macromolecules* **44**, 1570 (2011).

[Wedler 07]   G. Wedler. *Lehrbuch der physikalischen Chemie*. Wiley-VCH, Weinheim (2007).

[Wendlandt 97]   W.W. Wendlandt, P.K. Gallagher. *Thermal Characterization of Polymeric Materials*. Academic Press, San Diego (1997).

[Wierenga 05]   P. Wierenga. *Basics of Macroscopic Porperties of Adsorbed Protein Layers formed at Air-Water Interfaces based on Molecular Parameters*. Thesis Wageningen University (2005).

[Wilhelm 98]   M. Wilhelm, D. Maring, H.W. Spiess. *Rheol. Acta* **37**, 399 (1998).

[Wilhelm 99]   M. Wilhelm, P. Reinheimer, M. Ortseifer. *Rheol. Acta* **38**, 349 (1999).

[Wilhelm 02]   M. Wilhelm. *Macromol. Mater. Eng.* **287**, 83 (2002).

[Windhab 05]   E.J. Windhab, M. Dressler, K. Feigl, P. Fischer, D. Megias-Alguacil. *Chem. Eng. Sci.* **60**, 2101 (2005).

[Winstein 54]   S. Winstein, E. Clippinger, A.H. Fainberg, G.C. Robinson. *J. Am. Chem. Soc.* **76**, 2597 (1954).

[Worsfold 78]   D.J. Worsfold, S. Bywater. *Macromolecules* **11**, 582 (1978).

Bibliography

[Wu 05]   C. Wu. Handbook of Size Exclusion Chromatography and related techniques. Marcel Dekker (2005).

[Yilgor 98]   I. Yilgor, E. Yilgor. *Polymer Bulletin* **40**, 525 (1998).

[Yu 02]   W. Yu, M. Bousmina, M. Grmela, C. Zhou. *J. Rheol.* **46**, 1401 (2002).

[Yu 03]   W. Yu, M. Bousmina. *J. Rheol.* **47**, 1011 (2003).

[Zhao 09]   J. Zhao, M.D. Ediger. *Macromolecules* **42**, 6778 (2009).

[Zinchenko 97]   A.Z. Zinchenko, A. Rother, R.H. Davis. *Phys. Fluids* **9**, 1493 (1997).

11. Appendix

11.1 Materials and Methods

11.1.1 Synthesis of PI-*b*-PDMS in Chapter 7

For the anionic polymerization a clean reaction vessel and uncontaminated reaction agents are necessary with respect to oxygen and water, to avoid termination processes and deactivation of the initiator. All reactions were performed in glass reactors which were connected to a high vacuum line. To remove the water attached at the glass surfaces, the glass equipment is heated up under vacuum with a heat gun at its highest degree (about 500 °C) for several minutes. Afterwards the glass reactors are purged with inert Argon gas, which is also dried over activated NaA-zeolites molecular sieves (3 Å), while it is cooling down to room temperature before the heating and cooling is repeated for another two times. Before the monomers were used, they have been distilled. Hexamethylcyclotrisiloxane (HMCTS), see Figure 8.5, is solid at room temperature and was, therefore, dissolved in toluene, which facilitated the purification and distillation process. This solution was stirred for at least 3 h over CaH_2 before it was vacuum distilled at 90 °C. As a second purification step dibutylmagnesium, Bu_2Mg (1 M in heptane), was added to the Schlenk flask under Argon counter flow, where HMCTS was distilled in from the first purification step. The heptane was evaporated under vacuum before the monomer HMCTS was distilled. The amount of Bu_2Mg was chosen carefully, because it is able to ring-open the monomer HMCTS. The amount of ring-opened HMCTS by excess Bu_2Mg was about 2 %. Isoprene (99 % purity from Aldrich, $T_b = 34\,°C$) was distilled over n-BuLi (2.2 M in hexane, which was evaporated before isoprene was added), where 1 mL solution was used for 10 mL monomer. The solvent toluene was refluxed over CaH_2 for at least 24 h before it was stirred over PSLi which was used also as indicator for the purity of the solvent, with its bright orange color. THF was stirred over sodium after pre-drying over CaH_2, which reacts with the moisture in the solvent. The addition of benzophenone to the presence of sodium in THF shows a deep purple color, which indicates the purity of the solvent. Before the solvents were directly distilled from the solvent container into the reactor for the polymerization, they were degassed to remove gasses like oxygen, which were dissolved in the liquid.

Due to the synthesis of a single diblock copolymer the complete synthesis conditions with the amount of reaction agents shall be given below. The amount of reactants and the desired molecular weights of each block were determined by equation (8.3). For PDMS a molecular weight of 5.6 kg/mol and for PI of about 3.8 kg/mol should be achieved to be below the entanglement molecular weight. These molecular weights were used as calculation basis. The initiator concentration is calculated from the following relation of polymerization degree to the ratio of monomer to initiator concentration and polymer to monomer concentration given for a stochiometric living polymerization mechanism [Hsieh 96]:

$$P_n = \frac{M_w(\text{Polymer})}{M(\text{Monomer})} \quad \text{and} \quad P_n = \frac{[M]}{[I]} \ . \tag{11.1}$$

For PDMS a polymerization degree of $P_n = \frac{5600\,\text{g/mol}}{222\,\text{g/mol}} \approx 25$ is calculated. For the synthesis 8 g of the monomer HMCTS are inserted, whereas a conversion of 75 % yields a momoner mass of 6 g HMCTS for further calculations. The monomer amount of substance is calculated to $[M] = \frac{6\,\text{g}}{222\,\text{g/mol}} = 0.027\,\text{mol}$. Therefore the amount of sec-BuLi, the initiator, is defined by equation (11.1) to $[I] = \frac{0.027}{25} = 1.08 \cdot 10^{-3}\,\text{mol}$. Due to a 2 mol/L concentrated solution of initiator in hexane, 0.54 mL are added to the monomer isoprene, which is synthesized as the first block. The amount of isoprene adheres to the desired M_w of 3.8 kg/mol of the block. P_n of polyisoprene is calculated to be $\frac{3800\,\text{g/mol}}{68\,\text{g/mol}} \approx 56$, which defines the necessary amount of substance of isoprene reacting with the $1.08 \cdot 10^{-3}$ mol initiator:

$$[M] = P_n \cdot [I] = 56 \cdot 1.08 \cdot 10^{-3} = 0.06\,\text{mol} \xrightarrow{68\,\text{g/mol}} 4.1\,\text{g} \xrightarrow{0.68\,\text{g/mL}} 6\,\text{mL} \ . \tag{11.2}$$

For the termination trimethylchlorosilane (TMS-Cl) with a ten fold volume with respect to the initiator is added and the block copolymer is finally precipitated in an excess of methanol. After a re-precipitation of the block copolymer in THF and methanol a yield of 8.22 g polymer is achieved, which is a conversion of 81 % with respect to the used monomer amount. The characterization of the block copolymer is given in Section 8.5

11.2 Instruments

11.2.1 Experimental setup of the rotational rheometers ARES G1, ARES G2 and the Discovery Hybrid Rheometer DHR2

Within this thesis two generation of strain controlled rheometer were used to measure LAOS. For measurements of the colloidal dispersions in Section 6 the first generation rheometer from TA Instruments, the ARES G1 is used. The Rheometer was equipped with an electromagnetically compensated dual range Force Rebalance transducer (100FRT), which is suitable for measuring torques in the boundaries from $2 \cdot 10^{-4}$mN/m to 10 mN/m as specified by the manufacturer. Additionally a water bath adjusts the temperature in the range of $5\,°C$ to $95\,°C$. With its high resolution motor, frequencies from 10^{-5} up to 500 rad/s with varying deformation amplitudes from 0.005 to 500 mrad can be applied. The setup consists of the actual rheometer control with serial cable as external electronic connection between PC and measurement device and an independent FT-Rheology data acquisition and processing via double shielded BNC-type cables. The analog raw data of the measurements are digitized with a 16-bit ADC. With this ADC card a maximum sampling rate of 100.000 Hz is detectable, which means 50 kHz for each channel strain and torque. This high sampling rate causes a very small time between consecutive data points if compared to the time scale of rheological experiments and ensures no loss of information while measuring the response of the sample. Oversampling yields an improved signal-to-noise ratio by acquiring the time data at the highest possible sampling rate and preaveraging them on the fly to reduce random noise, see also Section 3.1 and in [van Dusschoten 01]. The 16-bit ADC card defines the discrimination steps and therefore the minimum detectable intensity of weak signals by its ability to discriminate the intensity of the signal [Skoog 92]. The recording and digitization is handled with a home written LabVIEW® software (National Instruments) and analysing with a home written MATLAB® routine.

The second generation of ARES G2 rheometers has the FT software as a commercial option implemented. Additionally a redundant external measurement is still possible as for the ARES G1. The motor is an air bearing supported, brushless DC motor, where position and rate are measured with an optical encoder including a position feed-back loop. The benefits of this motor are a non-invasive motor current, inertia and friction on sinusoidal strain controlled rheological measurements [Franck 08]. The rheometer consists of a Force Rebalance Transducer (FRT), which is suitable for measuring torques between 50 nNm and 200 mNm as specified by the manufacturer. The installed brushless DC motor with jewelled air bearings is capable of applying rotational angular velocities from 10^{-6} rad/s to 300 rad/s and deformation amplitudes of 1 μrad to an unlimited maximum. The applied oscillation frequency can be varied between 10^{-6} rad/s and 628 rad/s. The axial force ranges from 0.001 N to 20 N. For the ARES G2 software TRIOS, there are two possible methods to analyze the LAOS response. The first one is a real time correlation to determine the magnitude and phase of the fundamental frequency as well as the contribution of odd higher harmonics as a function of experimental time. A second possibility is the post processing of the acquired raw strain and stress data in the time domain detected via TRIOS software using a discrete Fourier Transformation with a specially written MATLAB® routine.

The third generation of rheometers, used within this thesis, is the new and just released Discovery Hybrid Rheometer DHR2 from TA Instruments, [TA Instruments 11]. There are three types of Hybrid Rheometers. Within this work the HR2 was used, which is equipped with a force rebalance transducer (FRT) capable to measure torques down to 2 nNm in oscillation. The FT software is as a commercial option implemented. The DHR2 is a stress controlled rheometer, but with the option of strain control. The motor consists of the patented drag cup design, with minimized inertia, temperature and friction influence. The bearing of the motor is designed by magnetic thrust bearing and of two radial bearings based on porous carbon plates, which allows oscillation frequencies between 10^{-6} rad/s and 628 rad/s and the very high torque resolution. The axial force sensitivity is 0.005 N and thus less sensitive than in the ARES G2.

11.2.2 Light Transmission Microscope

The Axiophot light microscope from Zeiss was used in transmission mode to determine the droplet size distributions of the polymer blends PDMS/PIB and PDMS/PI in Section 7.7.1. The objectives of the microscope were a Plan objective with a 40 fold magnification and an Epiplan objective with 20 fold magnification used with an ocular of 10 fold magnification. The magnification was chosen as a function of the droplet sizes. Radius distributions with a volume average radius below 5 μm was determined with the Plan 40x. All other distributions of the investigated emulsions with higher volume average radius were measured with the Epiplan 20x. For each sample 500 to 600 droplets were taken manually into account to achieve a proper description and adequate statistics.

11.2.3 Shear cell equipment for the microscope

The light microscope from Zeiss can be equipped with a Linkam shear cell CSS450. The sample area has a diameter of 30 mm. The gap can be chosen between 5 and 2500 μm. The micro-stepped motor allows 25 000 steps/revolution in oscillatory, step and steady mode of operation. Within this thesis the shear cell was used at room temperature, but allows a temperature variation between -50 to 450 °C.

11.2.4 SEC with RI detector

The SEC instrument consisted of an Agilent 1100 pump and an Agilent 1200 Differential Refractive Index (DRI) with two PSS SDV Lux 8x300 mm columns (103 and 105 Å pore size) from Polymer Standards Service GmbH (PSS), Mainz, Germany. The homopolymer PI and the block copolymer PI-*b*-PDMS were analyzed in THF at 25 °C and a flow rate of 1 mL/min.

11.2.5 Size exclusion cromatography with an additional IR detector

The SEC instrument consisted of an Agilent 1200 pump, an Agilent 1200 Differential Refractive Index (DRI) and a Fourier Transform infrared spectroscopy detector Vertex 70 from Bruker with two PSS SDV Lux 20x300 mm columns (linear S) from Polymer Standards Service GmbH (PSS), Mainz, Germany. The IR measurement is conducted in a transmission cell. The spectrometer is equipped with a potassium bromide (KBr)-beam splitter and a mercury cadmium detector (MCT). The block copolymer PI-*b*-PDMS was analyzed in THF at 25 °C and a flow rate of 1 mL/min. In a post processing a mathematical solvent suppression is applied to the recorded signal.

11.2.6 Differential scanning calorimetry

DSC measurements were performed on a Mettler Toledo DSC 30 in a temperature range from -160 to 25 °C. A heating rate of 2, 10 and 20 K/min under nitrogen gas atmosphere was used.

11.2.7 Two dimensional small angle X-ray scattering

The 2D SAXS instrument consists of a Hecus S3-Micro X-ray system with a point microfocus source and 2D X-ray mirrors. The scatteirng pattern was recorded with a two dimensional CCD Detector from Photonic Science. Additionally a block collimator systems was used to ensure low back ground scattering. This setup can measure $\underline{q}$ values from 0.06 to 4.7 nm^{-1}. Due to the use of a block collimator only the half 2D SAXS pattern can be recorded.

11.3 Technical drawings

Plate plate geometry

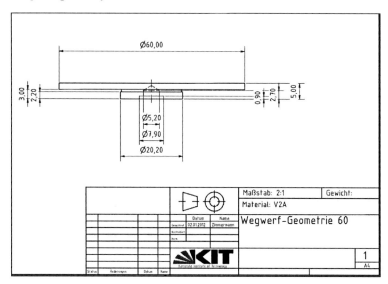

Figure 11.1: Technical drawing of the custom plate plate geometry with 60 mm diameter to be used with a commercial holder for disposable geometries of the ARES G2 from TA Instruments.

Sample holder of the Wilhelmy plate

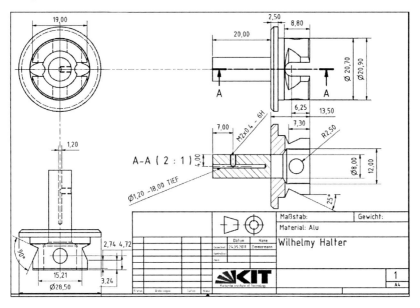

Figure 11.2: Technical drawing of the sample holder of the Wilhelmy plate used with the ARES G2 from TA Instruments.

Acknowledgements

Am Ende meiner Arbeit möchte ich mich bei denjenigen bedanken, ohne die, diese Arbeit nicht hätte stattfinden können und in dieser Form niemals abgeschlossen worden wäre. In erster Linie möchte ich mich bei Herrn *Professor Dr. Manfred Wilhelm* für die Möglichkeit bedanken, meine Doktorarbeit in seinem Arbeitskreis zu erstellen. Sie hatten die Gabe Interesse an neuen Themen zu wecken, den eigenen Kopf zum Denken anzuregen und motivierende Anstöße zu geben. Die wissenschaftlichen Gespräche konnten sich schon mal in die Länge ziehen, dafür war man spätestens nach einem Tag guten Mutes, dass es mit den Ergebnissen nur bergauf gehen kann. Außerdem möchte ich mich bedanken, dass ich in der glücklichen Lage war, meine Forschung mit einigen praxisnahen Produkten zu verbinden und meine gewonnenen Ergebnisse darauf anwenden zu können. Nicht zuletzt will ich hervorheben, dass ich mit Ihrer Hilfe viele interessante Konferenzen besuchen und einen unvergesslichen Aufenthalt in Sardinien erleben durfte.

Vielen Dank...

an die *Landesgraduiertenförderung* für das Promotionsstipendium.

an die *Deutsche Forschungsgemeinschaft* für die finanzielle Unterstützung: DFG SPP 1273 Kolloidverfahrenstechnik.

an die *Firma Beiersdorf* für die materielle und finanzielle Unterstützung und besonders an *Herrn Hetzel* und an *Dr. Brummer* für die netten und hilfreichen Gesrpräche.

Joseph Brader und *Miriam Siebenbürger* für die erfolgreiche Zusammenarbeit und eure nette Unterstützung, wenn ich auf dem Gebiet der MCT mal nicht direkt mitgekommen bin.

Massimiliano Grosso Grazie mille! Vi ringrazio per la cari ospitalita, un indimenticabile soggiorno a Cagliari, una paziente spiegazione la teoria e l'introduzione di nuovi software.

Herr Dingenouts für Ihre geduldigen und ausdauernden Korrekturen von Anträgen, Artikeln und wieder Anträgen und letztendlich dieser Arbeit.

der *gesamten Mannschaft der Lebensmittelverfahrenstechnik,* besonders Frederik Wolf und Lydia Schütz für die freundliche Einführung in die Geräte und die ausführlichen Diskussionen.

den *Rheologie-Kollegen vom MVM* für die Übernahme des alten ARES und das Willkommen sein an eurem Institut.

Ivanka Koleva für die vielen Messungen der Grenzflächenspannungen und den interessanten Tag an deinem Institut.

Frau Hörig und *Herr Arbogast* für die Messungen, die ich ohne euch nicht hätte machen können.

Daniel Zimmermann für die persönliche Betreuung meines Computers mit seinen kleinen und großen unerwarteten Aussetzern.

Jennifer Kübel für die interessante und ergiebige Zusammenarbeit und hilfreiche Unterstützung bei der Bierrheologie.

allen *Kollegen des Arbeitskreises Wilhelm,* die dafür gesorgt haben, dass das Arbeiten meistens eine Freude war und auf deren Unterstützung und Hilfsbereitschaft ich schon oft bauen konnte. Es war sehr hilfreich auf eure Expertisen im Bereich Synthese, Charakterisierung, Photographie und technischer Zeichnungen zählen zu können. Ich konnte viel von euch lernen und werde das meistens entspannte Arbeitsklima in guter Erinnerung behalten.

Meinen Bürokollegen möchte ich gerne danken, dass sie jederzeit Rücksicht auf meine Frostbeulen genommen haben.

Ein besonderer Dank gilt Christopher, Ingo, Michael und Thomas. Ihr seid während der Promotion zu liebgewonnen Menschen geworden und ich denke gerne an die lustigen Weihnachtsfeiern, Skiseminare, Grillfeste, Geburtstage, Urlaube, ... mit euch zurück.

Raphael, David und *Matthias* für ein meist angenehmes und lustiges gemeinsames Studium und den vielen hilfreichen Lernnachmittagen.

meinen Freundinnen *Maria, Julia, Helena* und *Julia* für die lustigen, gemütlichen, erholsamen, Neuigkeiten-austauschenden, feucht-fröhlichen und tröstenden Telefonate, Bingo-Nachmittage, Abende oder auch Wochenenden und Urlaube. Es ist schön zu wissen, dass räumliche Entfernungen unserer Freundschaft nichts anhaben und ein beruhigendes Gefühl, dass ihr da seid, wenn ich nach Hause komme.

meinen Eltern, denen ich den allergrößten Dank schulde, denn ihr habt mir ein sorgloses Studium ermöglicht und seid zusammen mit *Sabine* und *Susanne* und *meinen Omas* eine liebevolle Familie zu der ich immer wieder gerne nach Hause fahre, um eine entspannte Zeit zu haben.

Michael, du bist ein großartiger bester Freund, Mitbewohner und vor allem Freund. Auch deinen Eltern sei gedankt, dass sie uns beide immer willkommen heißen.

Curriculum vitae

Private data Kathrin Reinheimer

Ostendstr. 3

76131 Karlsruhe

Tel.: 01 51/ 17 00 43 49

E-Mail: kathrin.reinheimer@gmx.de

Born 08th of August 1984 in Wittlich (Rheinland Pfalz, Germany)

Education

08/1990–06/1994 Primary School, Wittlich

08/1994–03/2003 Peter-Wust-Gymnasium, Wittlich

Study

10/2003–10/2005 Prediploma in Chemistry at the University of Karlsruhe (TH)

10/2005–08/2008 Diploma in Chemistry at the University of Karlsruhe (TH)

February 2008 diploma thesis in the group of Prof. Dr. Manfred Wilhelm about *Fourier Transformation Rheology of Emulsions* at the University of Karlsruhe (TH).

10/2008–02/2012 PhD study at the group of Prof. Dr. Manfred Wilhelm, Institute for Technical Chemistry and Polymer Chemistry, Polymeric Materials

Stay abroad

04/2010–07/2010   Università degli Studi di Cagliari, Dipartimento di Ingegneria Chimica e Materiali, Piazza d'Armi, Cagliari, Italy in the group of Prof. Dr. Massimiliano Grosso

Publications

2010   J. Brader, M. Siebenbürger, M. Ballauff, K. Reinheimer, M. Wilhelm, S.J. Frey, F. Weysser, M. Fuchs, *Nonlinear response of dense colloidal suspensions under oscillatory shear: Mode-coupling theory and FT-rheology experiments*, Phys. Rev. E **82** 061401-20 (2010).

2011   K. Reinheimer, M. Grosso, M. Wilhelm, *Fourier Transform Rheology as a universal non-linear mechanical characterization of dilute monodisperse emulsions to investigate interfacial tension and size*, J. Colloid Interf. Sci. **360** 818-825 (2011).

             K. Reinheimer, M. Grosso, J. Kübel, F. Hetzel, M. Wilhelm, *Fourier Transform Rheology as an innovative morphological characterization technique for the emulsion volume average radius and its distribution*, J. Colloid Interf. Sci. (2011) accepted.

2012   K. Reinheimer, J. Kübel, M. Wilhelm, *Optimizing the sensitivity of FT-Rheology to quantify and differentiate for the first time the nonlinear mechanical response of dispersed beer foams of light and dark beer*, submitted to Z. Phys. Chem. (2012).

Conferences

ISFRS 2009   C. Klein, K. Reinheimer, M. Wilhelm, *Mechanical analysis of model emulsions and estimation of the particle size both base on FT-Rheology*, 5th International Symposium on Food Rheology and Structure (2009).

DRG 2010   K. Reinheimer, M. Grosso, P.L. Maffettone, M. Wilhelm, *Characterisation of high technology emulsions through advanced nonlinear mechan-*

ical spectroscopic measurements, Tagung der Deutschen Rheologischen Gesellschaft (2010).

SSR 2010   K. Reinheimer, *Fourier Transform Rheology of Emulsions*, Romanian Society of Rheology, 1st SSR - Summer School of Rheology (2010).

DKG 2010   K. Reinheimer, M. Grosso, P.L. Maffettone, M. Wilhelm, *Fourier Transform Rheology: A new approach to probe industrial emulsions*, Deutsche Kosmetische Gesellschaft Rheologie Workshop "Rheologie kosmetischer Emulsionen" (2010).

DRG 2011   K. Reinheimer, M. Grosso, M. Wilhelm, *Fourier Transform Rheology of Emulsions*, Tagung der Deutschen Rheologischen Gesellschaft (2011).

AERC 2011   K. Reinheimer, M. Grosso, M. Wilhelm, *Fourier Transform Rheology - a universal nonlinear mechanical characterization of dilute emulsions*, Annual European Rheology Conference (2011).

ECIS 2011   K. Reinheimer, M. Grosso, M. Wilhelm, *FT-Rheology a universal nonlinear mechanical characterization of polydisperse emulsions*, European Colloid and Interface Society (2011).

Courses

05/–07/03/2008   Dispersionen und Emulsionen Rheologie und Partikelgrößenbestimmung

31/08/–04/09/2009   Application of Neutrons and Synchrotron Radiation in Engineering Materials Science

Grants

11/2008–10/2010   Stipend of Landesgraduiertenförderung of Baden-Württemberg

04/2010–07/2010   Stipend of Deutscher Akademischer Austauschdienst (DAAD)

09/2011–present   Promotion of the Deutsche Forschungsgemeinschaft within the program of emphasis SPP1273 Kolloidverfahrenstechnik of the project

Fourier Transformations Rheologie als neue nichtlineare mechanische Charakterisierungsmethode für Emulsionen und Suspensionen auch mittels simultaner Kleinwinkellichtstreuung und NMR Charakterisierung

Awards

08/2009   Procter & Gamble prize for best diploma student in chemistry at the University Karlsruhe (TH) in 2008

Karlsruhe, April 4th 2012